Collaborative Embodied Performance

Performance and Science: Interdisciplinary Dialogues explores the interactions between science and performance, providing readers with a unique guide to current practices and research in this fast-expanding field. Through shared themes and case studies, the series offers rigorous vocabularies and methods for empirical studies of performance, with each volume involving collaboration between performance scholars, practitioners and scientists. The series encompasses the modalities of performance to include drama, dance and music.

SERIES EDITORS
John Lutterbie, Chair of the Department of
Art and of Theatre Arts at
Stony Brook University, USA
Nicola Shaughnessy, Professor of Performance
at the University of Kent, UK

IN THE SAME SERIES
Affective Performance and Cognitive Science
edited by Nicola Shaughnessy
ISBN 978-1-4081-8398-4

*An Introduction to Theatre, Performance
and the Cognitive Sciences*
John Lutterbie
ISBN 978-1-4742-5704-6

Performance and the Medical Body
edited by Alex Mermikides and Gianna Bouchard
ISBN 978-1-4725-7078-9

Performance, Medicine and the Human
Alex Mermikides
ISBN 978-1-3500-2215-7

Performing Psychologies
Edited by Nicola Shaughnessy and Philip Barnard
ISBN 978-1-4742-6085-5

Performing Specimens: Contemporary Performance and Biomedical Display
Gianna Bouchard
ISBN 978-1-3500-3567-6

Performing the Remembered Present: The Cognition of Memory in Dance, Theatre and Music
Edited by Pil Hansen with Bettina Bläsing
ISBN 978-1-4742-8471-4

Theatre and Cognitive Neuroscience
Edited by Clelia Falletti, Gabriele Sofia and Victor Jacono
ISBN 978-1-4725-8478-6

Theatre, Performance and Cognition: Languages, Bodies and Ecologies
Edited by Rhonda Blair and Amy Cook
ISBN 978-1-4725-9179-1

Collaborative Embodied Performance

Ecologies of Skill

Edited by
Kath Bicknell and John Sutton

METHUEN DRAMA
Bloomsbury Publishing Plc
50 Bedford Square, London, WC1B 3DP, UK
1385 Broadway, New York, NY 10018, USA
29 Earlsfort Terrace, Dublin 2, Ireland

BLOOMSBURY, METHUEN DRAMA and the Methuen Drama logo are
trademarks of Bloomsbury Publishing Plc

First published in Great Britain 2022
This paperback edition published 2023

Copyright © Kath Bicknell, John Sutton and contributors, 2022, 2023

Kath Bicknell and John Sutton have asserted their right under the Copyright,
Designs and Patents Act, 1988, to be identified as Editors of this work.

For legal purposes the Preface and Acknowledgements on pp. xviii–xxiii
constitute an extension of this copyright page.

Series design by Louise Dugdale
Cover image: *AURA NOX ANIMA* (© Lux Eterna)

All rights reserved. No part of this publication may be reproduced or transmitted
in any form or by any means, electronic or mechanical, including photocopying,
recording, or any information storage or retrieval system, without prior
permission in writing from the publishers.

Bloomsbury Publishing Plc does not have any control over, or responsibility for,
any third-party websites referred to or in this book. All internet addresses
given in this book were correct at the time of going to press. The author and
publisher regret any inconvenience caused if addresses have changed or sites
have ceased to exist, but can accept no responsibility for any such changes.

A catalogue record for this book is available from the British Library.

Library of Congress Cataloging-in-Publication Data
Names: Bicknell, Kath, editor of compilation. |
Sutton, John, 1965- editor of compilation.
Title: Collaborative embodied performance : ecologies of skill /
edited by Kath Bicknell and John Sutton.
Description: London ; New York : Methuen Drama, 2022. |
Series: Performance and science: interdisciplinary dialogues |
Includes bibliographical references and index. |
Identifiers: LCCN 2021032676 (print) | LCCN 2021032677 (ebook) |
ISBN 9781350197695 (hardback) | ISBN 9781350197756 (paperback) |
ISBN 9781350197718 (ebook) | ISBN 9781350197701 (epub)
Subjects: LCSH: Movement (Acting) | Movement (Philosophy) | Performance.
Classification: LCC PN2071.M6 C65 2022 (print) | LCC PN2071.M6 (ebook) |
DDC 792.02/8–dc23/eng/20211014
LC record available at https://lccn.loc.gov/2021032676
LC ebook record available at https://lccn.loc.gov/2021032677

ISBN:	HB:	978-1-3501-9769-5
	PB:	978-1-3501-9775-6
	ePDF:	978-1-3501-9771-8
	eBook:	978-1-3501-9770-1

Series: Performance and Science: Interdisciplinary Dialogues

Typeset by Integra Software Services Pvt. Ltd.

To find out more about our authors and books visit www.bloomsbury.com
and sign up for our newsletters.

CONTENTS

List of Figures x
List of Contributors xi
Preface and Acknowledgements xviii

Introduction: The situated intelligence of collaborative skills 1
John Sutton and Kath Bicknell

Part 1 Complex ecologies of embodied collaboration

1. Dropping like flies: Skilled coordination and Front-of-House at Shakespeare's Globe 21
   Evelyn B. Tribble

2. On the edge of undoing: Ecologies of agency in Body Weather 35
   Sarah Pini

3. A conversation on collaborative embodied engagement in making art and architecture: Going beyond the divide between 'lower' and 'higher' cognition 53
   Janno Martens, Ronald Rietveld and Erik Rietveld

Commentary: Redirecting our telescope 69
Amy Cook

Part 2 Learning, collaboration and socially scaffolded cognition

4 'No elephants today!' Recurrent experiences of failure while learning a movement practice 77
Kath Bicknell and Kristina Brümmer

5 Not breathing together: The collaborative development of expert apnoea 93
Greg Downey

6 Cultivating one's skills through the experienced other in aikido 109
Susanne Ravn

7 Musical agency and collaboration in the digital age 125
Tom Roberts and Joel Krueger

Commentary: Embodied learning within embodied communities 141
Emily S. Cross

Part 3 Symmetry and synergy in embodied coordination

8 Symmetries of social performance-environment systems 153
Rachel W. Kallen, Margaret Catherine Macpherson, Lynden K. Miles and Michael J. Richardson

9 Sing's trap: Staging low-commitment strategizing in muay thai 171
Sara Kim Hjortborg

10 Intercorporeal synergy practices – perspectives from expert interaction 187
Michael Kimmel and Stefan Schneider

Commentary: Mixing methods in the study of human action 207
Anthony Chemero

Afterwords

Commentary: Ecologies of acting and enacting 215
Catherine J. Stevens

Commentary: Betwixt and between 223
Ian Maxwell

Index 230

FIGURES

1. A snapshot from HD digital video, *AURA NOX ANIMA*. © Lux Eterna 40
2. RAAAF | Atelier de Lyon – Deltawerk // (2018). Photo by Jan Kempenaers. © RAAAF 59
3. RAAAF – Still Life (2019). Photo by Jan Kempenaers. © RAAAF 60
4. RAAAF | Atelier de Lyon – Sandblock (2019). © RAAAF 61
5. RAAAF | Atelier de Lyon – Trusted Strangers | New Amsterdam Park (N.A.P.). © RAAAF 63
6. Sensei practising the technique *kotagaishi* with Ito-san in Reishin Dojo. Photo by famenext (@famenext) 113
7. Illustration of Ecological Dynamics Approach, Symmetry Principles and Group Theory. © Michael Richardson 155
8. Acroyoga counterbalance (a) aligned 'stacking'; (b) opening a triangle with a pull connection. Photos by M. Kimmel 192
9. Push-hands: (a) 1–4 'push-yield', (b) 5–7 'retreat-pluck'. Photos courtesy of Loni Liebermann and Hella Ebel 197
10. Contact Improvisation lift. Photos by M. Kimmel 200

CONTRIBUTORS

Kath Bicknell's research investigates the relations between thinking, doing, performance and training. She is endlessly curious about how people make sense of skilful experiences as they happen, especially in time-sensitive scenarios like racing bikes down rock gardens in the jungle or performing at height on the trapeze. With research and teaching experience in Performance Studies (University of Sydney and the National Institute of Dramatic Art, Australia), Cognitive Science (Macquarie University, Australia) and Anthropology (Macquarie University) her work bridges the humanities and the sciences. Kath has worked as a freelance media professional since 2008 and is internationally recognized for her multiplatform work on cycling.

Kristina Brümmer is a Research Assistant at the Institute of Sport Science at the Carl von Ossietzky University of Oldenburg, where she obtained her PhD in 2014, and currently acts as a Substitute Professor for the Social Sciences of Sport at the Goethe University of Frankfurt/Main. Drawing on sociological practice theories, theories of the body and concepts of tacit knowledge, her research studies processes of subjectivation and team coordination in high-performance youth sport by means of qualitative research methods, especially ethnography, videography and interviews.

Anthony Chemero is Distinguished Research Professor of Philosophy and Psychology at the University of Cincinnati, USA. He is the author of more than 100 articles and the books *Radical Embodied Cognitive Science* (2009) and, with Stephan Käufer, *Phenomenology: An Introduction* (2015; 2nd edition, 2021).

Amy Cook is an Associate Dean for Research and Innovation and Professor of English at Stony Brook University, USA. She specializes in the intersection of cognitive science and theories of performance with particular attention to Shakespeare. She has published *Shakespearean Futures: Casting the Bodies of Tomorrow on Shakespeare's Stages Today* (2020), *Building Character: The Art and Science of Casting* (2018), *Shakespearean Neuroplay: Reinvigorating the Study of Dramatic Texts and Performance Through Cognitive Science* (2010) and co-edited *Theatre, Performance and Cognition: Languages, Bodies and Ecologies* (2016).

Emily S. Cross holds positions as a Professor of Human Neuroscience within the Department of Cognitive Science at Macquarie University, Australia, and as a Professor of Social Robotics within the Institute of Neuroscience and Psychology at the University of Glasgow, Scotland. Her background in performing arts has shaped her interest in action learning, neuroaesthetics and social robotics. Through her research, she uses a variety of methods (including brain-imaging techniques, action training paradigms and human–robot interaction) to explore how experience-dependent plasticity and expertise is manifest across brain and behaviour.

Greg Downey is Professor of Anthropology in the Macquarie School of Social Sciences at Macquarie University, Australia. He is author of *Learning Capoeira: Lessons in Cunning from an Afro-Brazilian Art* (2005) and editor with Daniel Lende of *The Encultured Brain: An Introduction to Neuroanthropology* (2012). Greg has done ethnographic research primarily in Brazil, Australia and the United States, and is currently the Executive Editor of *Ethos*, the journal of the Society for Psychological Anthropology.

Sara Kim Hjortborg is a PhD candidate in the Department of Cognitive Science, Macquarie University, Australia. She holds a Master's degree in Sports and Health from the Department of Sports Science and Clinical Biomechanics, University of Southern Denmark. Her research interests span martial arts, ethnography, skill and expertise, distributed cognition, enactivism, phenomenology and bodily self-awareness.

Rachel Kallen is an Associate Professor in the Department of Psychology at Macquarie University, Australia. Her research utilizes a framework of complex systems to investigate a range of social behaviours and systems. Bridging both basic and applied science, she has expertise in many areas of social psychology (i.e. stigma, intergroup relations, the science of diversity) as well as in cognitive science (social interaction, multiagent coordination and dynamical modelling).

Michael Kimmel is a researcher at the University of Vienna, Austria, with a focus on embodied, enactive, embedded and extended cognitive science. His range of topics includes interaction and joint improvisation, co-creation, embodied decision-making, skill theory and movement expertise, as well as expertise for complexity regulation. In addition to some biomechanical work, he has developed micro-genetic interview tools for reconstructing tacit and embodied knowledge (empirical phenomenology, stimulated recall, experimental workshops). Application fields include various forms of improvisational partner dance, martial arts, somatic therapy and partner acrobatics. Until 2013 he also worked on metaphor, imagery, sociocultural embodiment and narrative within a cognitive linguistics context, the field in which he took his PhD in 2002.

Joel Krueger is a Senior Lecturer in Philosophy at the University of Exeter, UK. He is part of the *Mind, Body and Culture* research group. He works primarily in phenomenology, philosophy of mind and philosophy of cognitive science: specifically, issues in 4E (embodied, embedded, enacted, extended) cognition, including emotions, social cognition and psychopathology. Sometimes he also writes about comparative philosophy and philosophy of music. Recent publications explore emotions online, as well as autism and the politics of everyday spaces.

Margaret Catherine Macpherson is a second-year PhD student in the School of Psychological Science at the University of Western Australia, currently working under the supervision of Dr Lynden Miles. Her doctoral work focuses on exploring the mechanisms underlying the relationship between interpersonal coordination and mental health. Her other research interests include applying dynamical systems approaches to the study of social interaction and examining the influence of synchronous action on prosocial and functional outcomes such as group rapport and productivity.

Janno Martens studied philosophy and architectural history at the University of Amsterdam, Netherlands. He is currently a PhD candidate at KU Leuven, Belgium, with a grant from the Research Foundation Flanders (FWO), investigating how technological and psychological notions of environment shaped architecture and urbanism in North America between 1965 and 1980. Before joining KU Leuven, Janno worked as research assistant for Erik Rietveld, partner at Rietveld Architecture-Art-Affordances (RAAAF), with whom he has co-authored several articles about the relation between architecture and ecological psychology. In 2018–19, he served as coordinator of the Jaap Bakema Study Centre in Rotterdam, which links the National Collection for Dutch Architecture and Urban Planning to the research agenda of the Architecture Department at TU Delft.

Ian Maxwell is currently Head of the School of Literature, Art, and Media in the Faculty of Arts and Social Sciences at the University of Sydney, Australia. His current research interests include the health implications of actors' lives, the history of romantic modernism in Australian theatre and phenomenologies of performance. His past work has included studies of youth cultures – particularly hip-hop practices in Australia – and the phenomenology of ritual practices. Prior to his academic career, Ian trained as a theatre director at the Victorian College of the Arts, Melbourne. His most recent theatre work was *Prince Bettliegend – A cabaret from Terezin*, which was based upon survivor testimony and various musical and textual traces from the ghetto city under Nazi occupation. The script created through that process, along with a series of essays by various collaborators in the project, is to be published in 2021, titled *Staging Trauma, Staging Pleasure*.

Lynden K. Miles is an experimental social psychologist in the School of Psychological Science at the University of Western Australia. He obtained his PhD from the University of Canterbury and has previously held academic positions in New Zealand and the UK. His research lies at the intersection of social psychology and complexity science, and spans both intra- and interpersonal aspects of social behaviour. An overarching theme of Lynden's research concerns the application of theory and methods consistent with an embodied-embedded approach to the study of social interaction, with a particular focus on interpersonal coordination.

Sarah Pini is Assistant Professor of Dance and Performance at the University of Southern Denmark, working across anthropology, phenomenology, performing arts, dance and cognition in skilled performance. Her research addresses notions of presence, embodiment and agency in different performance practices and cultural contexts. Alongside her academic research, Sarah's artistic practice investigates interconnections of movement, emotion and environment and how their dynamical relationships shape embodied narratives and sense making. Sarah's work has received several acknowledgements and has been showcased internationally. Her research appears in *The Oxford Handbook of Contemporary Ballet*, *Performance Research*, *Frontiers in Psychology*, and *The Journal of Embodied Research*, among other publications.

Susanne Ravn is Professor and Head of the Research Unit, *Movement, Culture and Society* at the Department of Sports Science and Biomechanics, University of Southern Denmark. She has published widely on the integration of phenomenology and qualitative research methodologies. She has been the leading investigator on several funded research projects focusing on dance practices, skilled performance in sport and health issues. Her research spans contributions to contemporary philosophical phenomenological discussions and more direct impact on cultural and societal issues. *Philosophy of Improvisation*, which she co-edited with Simon Høffding and James McGuirk, was published in 2021.

Michael Richardson is Professor of Psychology at Macquarie University, Australia. He is an experimental psychologist and cognitive scientist, with expertise in embodied cognition, social and perception-action psychology, complex systems, nonlinear dynamics, interactive virtual reality, human–machine interaction and AI. His research is directed towards identifying and modelling the dynamical processes that underlie human and multiagent perception, action and cognition, and the degree to which human-inspired dynamical and computational models can be employed to develop robust human–machine and human–AI systems.

Erik Rietveld is a Socrates Professor in Philosophy at the University of Twente and the University of Amsterdam (Amsterdam UMC, Dept of Psychiatry/Philosophy), Netherlands. Earlier he was a Fellow in Philosophy at Harvard University. He works on the philosophy of skilled action, change-ability and ecological psychology. Rietveld has been awarded an ERC Starting Grant and VENI, VIDI and VICI grants by the Netherlands Organisation for Scientific Research (NWO). Together with his brother Ronald Rietveld he founded the multidisciplinary collective for visual art, experimental architecture and philosophy RAAAF in 2006. They were responsible for *Vacant NL*, the successful Dutch contribution to the Venice Architecture Biennale 2010. He is a member of The Society of Arts of The Royal Netherlands Academy of Arts and Sciences (KNAW).

Ronald Rietveld graduated in 2004 *cum laude* at the Amsterdam Academy of Arts. His working period during Prix de Rome 2006 at the Rijksakademie of Visual Arts in Amsterdam was the early beginning of RAAAF. After winning the golden medal he founded this multidisciplinary and experimental studio together with his brother and Socrates Professor in Philosophy, Erik Rietveld. RAAAF works at the intersection of visual art, architecture and philosophy. RAAAF's work has been published worldwide and exhibited at leading contemporary art and architecture biennales such as those of São Paulo, Istanbul, Chicago and Venice. Ronald is a member of the Society of Arts of the Royal Netherlands Academy of Arts and Sciences (KNAW).

Tom Roberts is a lecturer in philosophy at the University of Exeter, UK, where he is a member of the *Mind, Body and Culture* research group. He works principally in the Philosophy of Mind and Psychology, on issues surrounding perception, emotion and their overlap. Recent topics on which he has published include how absences are encountered in emotions such as loneliness, how to understand the perceptual experience of awful noises and how aesthetic engagement with objects might be mediated by the sense of haptic touch.

Stefan Schneider is a cognitive scientist interested in embodied cognition, neuromotor pedagogy and movement practices. Currently, he is part of a research project on intercorporeal synergies (PI: Michael Kimmel). His PhD project at the University of Osnabrück, Germany, focuses on the role of body awareness for movement learning across somatic practices such as tai chi, Feldenkrais, gaga dance and ideokinesis. Stefan also holds a degree in fine arts and is a certified tai chi teacher.

Catherine (Kate) Stevens, a cognitive scientist, is Director of the MARCS Institute for Brain, Behaviour & Development at Western Sydney University, Australia. She holds BA (Hons) and PhD degrees from the University of Sydney. Kate conducts basic and applied research into the learning, perception, creation and cognition of complex actions. She also applies methods from experimental psychology to investigate human–machine interaction (e.g. design of auditory warnings, human–avatar and human–robot interaction). Kate is author of more than 200 peer-reviewed papers. She is Editor-in-Chief, *Music Perception*, Professor in Psychology and Pro Vice-Chancellor STEM at Western Sydney University.

John Sutton's research addresses memory and skill, across the cognitive sciences and the humanities. He is the author of *Philosophy and Memory Traces: Descartes to Connectionism* (1998), and he has coedited three previous books – *Descartes' Natural Philosophy*, *Embodied Cognition in Shakespeare's Theatre* and *Collaborative Remembering*. He seeks to integrate conceptual, experimental and ethnographic methods, and has published on memory and skill not only in philosophy and cognitive science, but also in archaeology, film, history, linguistics, literature, music, psychology and sport science. He is a Fellow of the Australian Academy of Humanities, and past President of the Australasian Society for Philosophy and Psychology.

Evelyn Tribble is Professor of English and Associate Dean at the University of Connecticut, USA, where she has been employed since 2018. From 2003 to 2018 she was Professor of English at the University of Otago, Dunedin, New Zealand. Her books include *Cognition in the Globe: Attention and Memory in Shakespeare's Theatre* (2011) and *Early Modern Actors and Shakespeare's Theatre: Thinking with the Body* (2017). She is currently working on the Arden 4 edition of *Merry Wives of Windsor*.

PREFACE AND ACKNOWLEDGEMENTS

Kath Bicknell and John Sutton

This book is about joint intelligence in action. In performance and the arts – dance, theatre, music, architecture – small groups of expert practitioners work together in rich, dynamic settings. In sport and martial arts, skilled individuals with distinct capacities coordinate or compete, constantly meshing with or responding to each other's movements. In collaborative performance, people act together – sometimes, for a time – as if of one mind.

How does tightly knit collaboration, in challenging skill worlds, work? How do some skilled performers enact their craft under tight time pressures, while giving the impression they have all the time in the world? How do some people seem to know in advance what to do *together* – how to move or respond in just the right ways even in volatile, stressful or rapidly changing environments? If these questions interest you, or prompt you to ask related questions about your own areas of interest, this book is for you.

In bringing authors together to write ten chapters on collaborative embodied skills, we hope to attract readers who are performers and fans, teachers and coaches, practitioners and critics, students and researchers. The book assumes no specific background in any one academic discipline, but draws on and seeks to contribute to many. Studying complex ecologies of skilled practice across distinctive, culturally unique environments and tapping the experience of highly trained specialists, our contributors examine the nature and mechanisms of collaborative performance in context.

Research groups too form their own unique cognitive ecologies. Individuals work together on projects which unfold over time in unpredictable ways, in changing and often challenging circumstances.

Any thoroughly interdisciplinary group, like the one we've been part of over the years, develops in unexpected directions as institutional, intellectual and interpersonal constraints shift. The hard, on-the-ground work of boundary-spanning research requires persistence, tolerance and slow-brewing trust. It requires enormous luck in finding the right collaborators, people who can cope and flourish even when, as they say, outside their comfort zone. We have had much of that luck. Integrative, sustained interdisciplinary research is much more difficult – more time-consuming, more draining, more resource-intensive, more fragile – than is acknowledged in glib management or policy documents about dissolving silo mentalities. But when it does go well it brings pleasure and surprise on many fronts.

This book presents diverse but coherent new work on collaborative embodied performance from a thoroughly international and interdisciplinary cohort of fellow travellers, of many different backgrounds and career stages. We and our contributors – seventeen authors of ten chapters, plus five commentators – are participants in distinctive ongoing conversations about collaborative embodied skills. We very much hope that you, as readers, will feel welcome to join these conversations and be inspired to contest or expand on the case studies showcased here. Our authors' affiliations span anthropology, architecture, cognitive science, dance, literature, neuroscience, performance studies, philosophy, psychology, sociology and sport science, but they are driven by their topics rather than any one tradition. We took to referring to this book project as 'stuff we like by people we like', which reveals the fun we have had bringing it to fruition.

John began research on movement skills in individual expertise, with Doris McIlwain, in 2004. Our early and ongoing debates and collaborative projects with Wayne Christensen (about cognitive control and automaticity), Greg Downey (about neuroanthropology and culture), Andrew Geeves (about music, emotions and performers' experiences) and Lyn Tribble (about skills in history) helped hugely as our group developed a theoretical focus on the idea of meshed control in performance, and a set of mixed methods to fuel what we called 'experience-near' case studies on skill in sport, yoga, dance and music.

Kath Bicknell joined our group in 2013, adding further expertise in ethnography and performance studies to the interdisciplinary mix. She forged her own pathway into this interdisciplinary world, strongly shaped by staff and students in the University of Sydney's Department of Theatre and Performance Studies where she studied and worked between 2003 and 2012. Paul Dwyer and Kate Rossmanith ignited a lifelong passion for ethnography. J. Lowell Lewis's provocative seminars on embodiment fuelled a thirst for exploring an often-felt disconnect between lived experiences and theoretical debates. Ian Maxwell's encouragement towards curiosity-driven research, and interdisciplinary approaches to doing it, continues in this volume's afterword. Working closely with John and what we now call the *Cognitive Ecologies Lab* at Macquarie University has provided fun and fertile ground from which to explore cognitive and performance theory through embodied practice, ongoing collaboration and the many joyful, excited discussions that come with being part of a deeply engaged, interdisciplinary team.

After many attempts and the usual bewildered frustration, as an evolving and growing group we were awarded funding for work on skilled performance from the Australian Research Council (ARC). We were lucky in that philosophy and the cognitive sciences alike had seen dramatic expansions in interest in expertise and skilled performance, with increasing integration of conceptual, experimental and ethnographic approaches. Our search for and exploration of rich middle ground between over-intellectualist and more 'mindless' approaches to individual skill was joined by more and more theorists. For more than fifteen years, Macquarie University provided solid support for our interdisciplinary research, both institutionally through the Department of Cognitive Science – Max Coltheart's glorious experiment – and CEPET, the University Research Centre for Elite Performance, Expertise, and Training, and collegially over many years of collaborations, joint activities and lively debates with our friends right across campus. Many students and visitors made vital contributions to challenge and to help sharpen our ideas. Our networks of allies, fellow travellers and critics continued to expand across disciplines and geographical locations. Despite other significant differences, we found many researchers sharing a fascinated commitment to thick, experience-near, immersive, practice-oriented, case-study-based approaches to performance.

It was becoming clear at this point that we needed to cast our net wider, to focus also on *collaborative* skills. The independent development of 4E (embodied, embedded, enactive and extended) approaches to cognition had provided encouragement and tools for studying social and ecological dimensions of performance, but had been mainly applied to other cognitive domains like memory, decision-making, emotion, language, navigation or tool use. The essays in this book expand the 'cognitive ecologies' framework to address skilled performance, with this specific extra focus on collaboration and joint action.

A number of our contributors gave talks at a workshop on collaborative embodied skills which John organized at Senate House in London in 2017, during a fellowship at the Institute of Philosophy. We were lucky to win a further ARC grant on this topic for 2018–21. The concrete plan for this book took shape in enjoyable conversations we had at, and after, the 2019 Cognitive Futures in the Arts and Humanities conference in Mainz, Germany. It was delightfully quick and easy to sign up our contributors. Indeed we soon realized how naturally and directly a second volume might follow: we warmly invite readers to let us know of other directions and ideas that might fuel that next step.

It has been a pleasure throughout to work with Methuen Drama, and we are thrilled to publish this book in the *Performance and Science* series. In offering both enthusiastic encouragement and well-informed critical input, series editors John Lutterbie and Nicola Shaughnessy have helped us greatly from the start. Lara Bateman, Mark Dudgeon, Ella Wilson and all at Methuen Drama have made the publishing processes smooth throughout. Our thanks also to Dharanivel Baskar and the production team at Integra. We are very grateful for all this assistance. Many thanks too to Lux Eterna for the evocative cover photo (for more about the related film, *AURA NOX ANIMA* (2016), see Chapter 2, by Sarah Pini).

The initial chapter drafts were written as we all endured lockdown in 2020. We ran three online workshops in August 2020 at which contributors presented work in progress across distinct time zones, generating feedback and cross-fertilization. At the next stage, each chapter draft was reviewed by two peers, in many cases by one other contributor and one external expert. Our five commentators – well-established skill researchers in theatre, robotics, philosophy, performance studies and psychology – played significant roles in these phases of project development, helping to sharpen all the chapters before completing the commentaries on resulting themes which you will read in the book. We have been pretty hands-on editors, engaging in iterative discussions with the contributors as each chapter went through multiple versions. It is our hope that despite the diversity of theories, concepts and approaches in the volume, this close interaction between editors, chapter authors and commentators has generated a genuine, and unusual, coherence of method and framework across diverse domains.

Our work on this book has been supported by ARC Discovery Project grants DP130100756 'Mindful Bodies in Action' and DP 180100107 'The Cognitive Ecologies of Collaborative Embodied Skills', for which we are very grateful. Such support for insistently interdisciplinary research is rare and precious. Doris McIlwain died in 2015, but her influence on this book is immense: we wish she could have participated. We are deeply appreciative of our entire team of authors and commentators, whose patience, commitment and sheer brio saw us through some challenging times and brought us the deep, long-term buzz and pleasure of real collaborative action. We want to single out Wayne Christensen, Andrew Geeves and McArthur Mingon for their help, their support and their vision at different stages of this research – thank you. Special thanks too to Greg Downey and Lyn Tribble for comments on our editors' introduction. We are also very grateful to the following other friends, colleagues, students, referees and reviewers, critics and collaborators who have directly encouraged and inspired, facilitated and contributed to our projects on skill along the way: Bruce Abernethy, Lucas Bietti, Max Cappuccio, Amanda Card, Andy Clark, J. M. Coetzee, Giovanna Colombetti, Ed Cooke, Rochelle Cox, Robin Dixon, Paul Dwyer, Matthew Elton, Regina Fabry, Damian Farrow, Ellen Fridland, Rasmus Gahrn-Andersen, Shaun Gallagher, Petra Gemeinboeck, Elle Geraghty, Celia Harris, Simon Høffding, Dan Hutto, Jesus Ilundain, Samuel Jones, David Kaplan, Nick Keene, Paul Keil, Carla Lever, J. Lowell Lewis, Julie-Anne Long, Ole Lund, Glen McGillivray, Jeremy McKenna, Clare MacMahon, Lambros Malafouris, David Mann, Lars Marstaller, Judith Martens, Paul Mason, Rich Masters, Richard Menary, Barbara Montero, Sean Müller, Rebecca Olive, Garth Paine, David Papineau, Carlotta Pavese, Karen Pearlman, Gert-Jan Pepping, Beth Preston, Ian Renshaw, Anina Rich, Dan Richardson, Kate Rossmanith, Justine Shih-Pearson, Tim Sinclair, Line Simonsen, Phil Slater, Barry Smith, Ben Smith, Kim Sterelny, Bill Thompson,

Anthony Uhlmann, Dave Ward, Glenn Warry, Alex Wessling, Mike Wheeler, Justine Whipper and Kellie Williamson.

Kath adds her personal thanks to Anne, Geoff, Brendan and Lauren Bicknell, Gaye Camm and her incredible extended family. And from John to Graeme Friedman, Christine Harris-Smyth and Nina McIlwain – thank you.

Introduction: The situated intelligence of collaborative skills

John Sutton and Kath Bicknell

The topic: Thinking with our feet

People move together, and do things together, all the time. We play and work and talk and suffer together, finding ease or joy, sharing pleasure or grief. We discover challenge, thrill and risk.

Joint actions may involve physical, manual or technical skill, and may rely on tools, technologies and ordinary old objects. Collaborative actions also involve *situated intelligence,* a dynamic, lively and social form of cognition. This book is a celebration and exploration of these things: the dizzying variety of remarkable ways that people move and think together, in unique places and settings, at a time and over time.

In initial orientation to the book's topics, we introduce in turn the five key concepts which animate it: performance, body, collaboration, cognition and ecology. We briefly describe the domains of **performance** in question here, its bodily or '**embodied**' nature, the forms of **collaboration** addressed, the role of intelligence or '**cognition**' in expert movement and the notion of '**ecologies** of skill'.

Performance

For practitioners and researchers in performance studies, we aim to do justice to the lived complexity of emotion, awareness and thought in skilled, coordinated action. What makes performance, sometimes, so precise, adaptable and marvellous? The collaborative activities we address span the full spectrum of performance, embodied practices and ecologies of skill: from aesthetic contexts such as theatre, architecture and music to sport and martial arts. Our primary focus is on specialist domains, in which skills must be laboriously acquired, and expertise is a gradual, fragile, wonderful outcome. We examine skilled tasks and activities operating at a range of nested and interacting timescales: from incredibly fast decision-making under severe pressure, to long-term shared histories of collaboration in rich cultures and subcultures.

Embodiment

Our contributors offer lively, animated accounts of the bodily and emotional nature of skilled performance, with many developing ethnographic or 'experience-near' case studies. Where theoretical work on 'embodied cognition' can be a little thin or abstract, here are vivid descriptions of striking bodily experiences as skills are honed and exercised. We find pain and visceral agony, and the screaming of muscles, as bodily capacities are stretched and remoulded in particular patterns of use. There is surprise and delight as experts and novices find new ways to move or coordinate. In concrete descriptions and analyses of diverse and specific bodily and emotional experiences, our contributors illuminate with precision the flexible intelligence that dancers, divers, fighters, composers, film directors and dedicated Front-of-House volunteers reveal in, and through, action.

Collaboration

In some cases, the forms of collaboration studied are dyadic, involving pairs of skilled actors – either cooperative or, as in martial arts, antagonistic. In other cases, collaboration is at the level of small groups or larger organizations. These are *social* units, composed of individual humans with unique characteristics, histories and skills, such that each member of a group brings something distinctive to the collaboration. The chapter by Roberts and Krueger highlights a human–machine system, in musician Holly Herndon's collaboration with an artificial neural network *Spawn*. All collaborating groups are situated in hybrid ecologies of their own: socio*material* systems incorporating technological or environmental resources. Collaborators rely on, deploy and create shared histories as essential components of their interactive skill. Such shared history can encompass emotions and emotional experiences, beliefs, worldviews and motivations, and intelligence or styles of decision-making. Shared pasts drive present performance.

Cognition

Skilled performers sometimes worry about overthinking. Practitioners and theorists alike, in sport and the arts, sometimes give the mind – cognition – bad press, fearing that too much thought may disrupt well-grooved actions or interfere with the body's smooth, instinctive responses. Teachers, coaches, critics and peers often support training regimes which encourage performers to 'keep it simple'. According to widespread lore, thinking is typically slow, effortful and clunky, a sign that something has gone wrong, or that performers are stuck in their heads, in a realm of inner deliberation disconnected from intuitive practice. Hubert Dreyfus, an influential philosopher of skill, wrote that 'mindedness is the enemy of embodied coping' (2007, 353). This book builds on recent criticism of this dominant approach, arguing that mind and cognition are not absent in embodied performance (e.g. Montero 2016; Sutton et al. 2011) to develop a very different picture of skilful action.

Our contributors describe in rich detail forms of thinking which are fast, dynamic, exquisitely adjusted to changing situations and deeply attuned to bodily capacities. In these contexts, thinking is public, right there in the shared world, expressed in or constituted by the expert performer's actions. In this book we reclaim, revel in and excavate the vital *cognitive* dimensions of skilled movement practices – *thinking* on and with our feet, and on the fly. We use the word 'cognition' in its broadest senses, not restricted to reasoning or to information processing, but to include the full diversity of embodied mental life: imagining, grieving, remembering, sensing, noticing, dreaming, wondering, listening, problem-solving, strategizing, pattern detecting and indeed designing, balancing or creating. In this capacious sense, cognition includes emotion and motivation, and is not located in the individual brain alone, no matter how important neural processes may be. Rather, the term signals flexible embodied intelligence, manifesting in experience and in action, in a social and material world.

Cognitive ecologies of skill

Cognitive processes, as embodied and revealed in skilled action, are richly integrated with and in the emotional, social, cultural, technological and technical dimensions of a performance setting. In this book we apply the idea of a 'cognitive ecology' to the specific domain of skilled performance, offering a new research focus on 'ecologies of skill'. The notion of a cognitive ecology 'points to the web of mutual dependence among the elements of an ecosystem' (Hutchins 2010, 706). When we dig down into the specific ecosystems within which embodied collaborations thrive, we find vast and uneven domains of resources and components. As well as the social interactions among collaborators and those around them, and the techniques and practices and strategies they have developed individually and collectively, we also find equipment – artefacts and technologies and devices, each with their own histories, their own formats, their own dynamics. Further, skilled performance occurs in diverse locations – places, specific sites, unique settings. Our contributors take you from the reconstructed Globe Theatre in London, to the sand dunes of Anna Bay in Eastern Australia, to gyms, studios and pools. These are *cognitive* ecologies, underlying or constituting the flexible embodied intelligence of skilled performance processes. The physical, environmental, technological and social resources are not merely external to the performers' mental lives, not just triggers or cues to action. Rather, our contributors show, their operations in interactive systems directly drive or sculpt what experts notice, do and decide. These rich ecologies scaffold and partly constitute the exquisite attention to salient changes that performers consistently exhibit. Such attention grounds their abilities to intuit or probe new action opportunities, or to rapidly shift tactics.

Building on this short overview of our topics, we discuss three orienting and integrative themes in more depth to highlight the distinctive contributions of this book: cognitive ecologies within the cognitive humanities, collaborative skills and methods for tapping situated intelligence. We conclude our introduction with a brief explanation of the structure of the volume, describing its sections and sharing a preview of each chapter.

Cognitive ecologies and the cognitive humanities

Cognitive approaches to skill and expertise are an increasingly significant component of performance studies. The cognitive humanities, more broadly, flourish. The success of Bloomsbury's well-established *Performance and Science* series, in which this book appears, as well as handbooks, overviews and many specialist works in the cognitive study of theatre, dance, music, film, literature and history, confirms the productive entanglement of the cognitive sciences and the study of performance and the arts (see, e.g. Blair and Cook 2016; Hart and McConachie 2010; Kemp and McConachie 2018; Lutterbie 2019; Pearlman, MacKay and Sutton 2018; Shaughnessy and Barnard 2020). The cognitive approaches to performance showcased and put to work in this book comprise a focussed and coherent subset of this interdisciplinary field. Most contributors apply and expand on one rich and increasingly mainstream tradition in cognitive theory, often called '4E cognition'. On this perspective, mental life is *embodied* rather than restricted to the brain alone, *embedded* in rich material and sociotechnical settings, *enacted* in histories of flexible engagement with the environment and sometimes *extended* across non-biological or cultural resources outside the body which may become integrated into hybrid cognitive systems. The '4E' label has its limitations, but is now well entrenched (Newen et al. 2018). It is deployed throughout this book, where the dynamic cognitive processes involved in collaborative performance are also seen as fundamentally *enculturated,* affective or *emotional* and intrinsically *ecological* in nature.

4E cognition emerged in the 1980s as a result of several interacting factors. Within the cognitive sciences, frustration at the failures and brittleness of classical AI systems gave rise to constructive alternatives across many fields including robotics, phenomenology, developmental psychology, connectionist modelling, dynamical systems theory and the 'extended mind' hypothesis in philosophy (Clark 1997; Clark and Chalmers 1998; Hurley 1998; Port and van Gelder 1995). The idea of distributed cognition – of cognition as potentially spread or 'distributed' across disparate physical and cultural resources – arose in parallel in the social sciences, integrating organizational and cultural psychology, science studies, archaeology, education, studies of human–computer interaction and especially cognitive anthropology (Hutchins 1995; on the history and on relations between these distinctive movements see Michaelian and Sutton 2013). To adapt the notion of 'cognitive ecologies' from biology and environmental science is to highlight the distinctive balances in operation across the varying components of interactive cognitive systems (Hutchins 2010; Johnson 2015; Smart et al. 2017; Tribble and Keene 2011). The cognitive ecologies framework directs us to investigate the shifting divisions of roles and labour across different people (with diverse capacities), artefacts and the physical and social structures of any complex environment.

Theories which treat intelligence as situated or distributed, and as a matter of active embodied practice, are natural allies for performance studies and the humanities. The 4E cognition movement in general, and the cognitive ecologies framework in particular, do not suffer from two flaws which have led many wise humanists to resist other influences from the cognitive sciences (compare Sutton and Keene 2017). First, the pervasive *individualism* of classical cognitivism, which located minds exclusively in the head or in a hidden subjective realm, is decisively rejected. Cognitive processes are not to be explained in reductionist fashion, by reference to the brain alone. Rather, these situated and ecological views take intelligence out of the head, finding it in embodied practices and in interactions 'with culturally organized material and social worlds' (Hutchins 2010, 712). So 4E-style cognitive humanities theory is not 'neurohistory': neural processes may be key components in larger hybrid cognitive systems, but are themselves deeply shaped by the cultural norms and practices in which they are integrated, for brains too are 'biosocial organs permeated by history' (Cowley 2002, 75).

Secondly, the cognitive ecologies framework is not universalizing or *imperialist*, concocted to engineer a scientific takeover of humanities research. Rather, the hope and expectation is for mutual benefit across fields. In one direction, the cognitive lens provides a new angle on real performance settings, and on independently motivated problems about collaborative skill. But then, as this book confirms, in reverse angle these analyses of performance sharpen, test, extend and develop the cognitive framework itself, so it can embrace rich real-world case studies of joint action in context. We hope that experimental psychologists, and others who study skill acquisition, motor learning or collaboration and joint action in laboratory settings, will also find much to interest them here, in qualitative, ethnographic and case study-based analyses which complement more controlled methods of enquiry into expert performance. Reciprocal insight of this kind is a tough but necessary goal for insistently interdisciplinary research. Rather than randomly sampling cognitive theories to import into performance studies, our contributors acknowledge the internal complexity of the relevant sciences, which are often deeply riven by tensions and disagreements.

Work on distributed cognitive ecologies affords a lively pluralism for performance studies and the cognitive humanities. This mode of interdisciplinary engagement elicits attitudes to cognitive theory that are subtler than either blunt critique or overconfident ebullience. Our contributors demonstrate the virtues of productive entanglement (compare Fitzgerald and Callard 2015; Tribble and Sutton 2014). They draw on the tools, concepts and methods their domains dictate, finding resources in theoretical debate and distinctive traditions of real-world practice. A number of chapters, for example, can be read as interventions in long-running controversies about the 'bounds' of cognitive systems – the question of whether a distributed or ecological approach is overly inclusive, generating a vague or 'unscientific motley' that threatens our grip on where human agents end and the rest of the world begins (Adams and Aizawa 2001, 62; see also Menary 2006; Sutton 2010). But our authors embrace motley from *within* their case studies, from within the worlds of practising architects, divers or fighters, not on abstract or detached theoretical grounds. In Tribble's dramatic expansion of the unit of analysis for theatrical performance, for example, the case for including Front-of-House stewards and the challenging management of audience behaviour in an account of theatrical performance is firmly anchored in the unique norms and expectations associated with the Globe Theatre as a unique ecology.

Like most boundary-spanning concepts, this notion of a cognitive ecology is not a magic bullet: it won't suddenly do our explanatory work for us, or shortcut the challenges of on-the-ground studies of creativity or performance. It does not allow us to bypass existing specialist research, or excuse us from homing in carefully on tiny and revealing facets of the richer webs. But our bet is that refinement, improvement and expansion of the theoretical framework will be most effectively achieved not by more grand metaphysical wrangling, but by building a repertoire of strong case studies. Together, such case studies illuminate distinctive aspects of the rich ecologies of collaborative embodied performance. Readers will identify novel ideas emerging from the interaction of the component studies.

The cognitive ecologies framework affords our contributors a shared focus within their broader commitment to 4E cognition. The chapters are united by two further layers of specificity, to which we now turn. Within the wider body of work on cognitive ecologies, we focus in on ecologies of *skill*, identifying embodied joint action and expert movement as an intriguing domain for probing the fundamentally *situated* nature of performance. Finally, among many ways of studying ecologies of skill, we share a commitment to thicker description and analysis of real-world collaborative performance.

Collaborative skills

Work on 4E cognition and on cognitive ecologies has influenced many fields and been applied to many domains, such as memory, moral reasoning and language. But these approaches have rarely been applied to performance or to skill. Leading theorists have invoked jazz improvisation, circus performance or team sport as metaphors or models for embodied cognition or intelligent action *in general*, but have often neglected further examination of the specifics of each kind of skill in its unique settings. We take the idea of 'ecologies of skill' beyond metaphor, to study the bodily and emotional experiences of experts and the interactive mechanisms of collaborative performance in context, at multiple timescales (Sutton and Bicknell 2020). In attending to the integration of cognitive, affective, social, technological and environmental aspects of skill acquisition and performance, this volume builds on recent cognitive approaches to the arts, as mentioned above, and sport (Cappuccio 2019).

On fundamental questions about the nature of skill and collaboration, these diverse case studies start from firm shared ground. Firstly, we treat skilled movement as deeply mindful and intelligent: expert performance is not automatic, not a set of fully proceduralized responses of experienced bodies simply triggered by current stimuli. Practitioners' talk of 'instinct' or 'intuition' is not the final word on the springs of action, but marks the complexity and dynamism of the multiple, meshed cognitive processes apparent in adaptive actions in context. When we dig into the real-time operations of skilled performance, where experts are seeking to improve and to go beyond their comfort zones, we find not simply smooth or hitch-free coping, but dizzying arrays of hard-won strategies and many forms of variable, free-floating awareness. In our own research group, for example, we have shown this in cricket, music, yoga, mountain biking and trapeze (Bicknell 2021; Christensen et al. 2015; Geeves et al. 2014; McIlwain and Sutton 2014; Sutton 2007; for similar findings from other perspectives see Fridland 2014; Toner et al. 2021). Cognition here does not lie *behind* skilful action, encoded as fixed control plans or transmitted as top–down instructions to an automated motor system. Rather, thinking is itself a bodily and worldly aspect of our real-time engagement with the rich resources of our shifting performance ecologies.

Secondly, in expanding such approaches to skill to encompass *joint* action, we treat collaboration as itself complex, taking many forms. In dyads, small groups and teams, individual performers bring distinct capacities to the collective. They each play their own specific part, dividing the work that progresses shared goals. To focus on collaborative interaction in embodied skills is not to efface or downplay the role of the individual (De Jaegher et al. 2016). It's because team members have different specialist skills that they complement each other, in favourable circumstances meshing to produce a performance that is 'greater than the sum of its parts'. Such collaborative or creative *emergence* (Sawyer and DeZutter 2009) is joint intelligence in action. Of course, success is precarious: in high performance especially, things can go wrong at any point. But in any particular case, we can examine what individuals *bring* to the collaborative skill; what individuals *do* in embodied performance; and how the cues to joint action are distributed across social and material cognitive ecologies. Fascination with the processes of teamwork and collaboration is a further shared starting point for this volume.

Fusing these themes about skill and collaboration, a recurrent topic across the chapters concerns the mechanisms of instruction in skill acquisition. It takes time to develop technically demanding action capacities, often in arduous, dedicated training regimes over years of laborious commitment. Given that we can't simply pick up a bodily skill all at once, as we might learn a fact, it's puzzling to understand how teachers and coaches influence performance, and how expert practitioners can influence themselves in practice to respond differently in changing situations. A number of contributions address the varieties of social and environmental 'scaffolding' or support for the learning and the exercise of embodied skills. Experienced teachers – in freediving or aikido, for example – instil and deploy cues or 'instructional nudges' that are finely tuned to the learners' current capacities, extending them to the next stage of development and gradually exposing them to the full cultural history of practice in the domain. Peers working together – in handstand classes, or in an architecture and design studio, or a recording studio – find novel response options emerging out of familiarity in interaction, with trust and ease built up and registered, for example, through shared linguistic or emotional motifs. In each case, instruction and support is also actively guided by the intelligence of the entire ecology: the equipment in a studio or a gym, or the norms and rituals of behaviour within an established movement discipline embody and transmit lessons from the history of practice in a domain (Hahn and Jordan 2014).

The real-time exercise of embodied skills continues to be reliant on forms of scaffolding. Even if some performers – like the expert muay thai fighter in Hjortborg's chapter – can cue or influence themselves in action, with less immediate dependence on teachers, real experts never stop seeking ways to improve. Yet much of the machinery of scaffolding, distributed as it is across complex ecologies, collaborative groups and whole systems in motion, can be nearly invisible. This raises significant problems of method, which we consider briefly before stepping aside to let our contributors respond.

Methods for tapping situated intelligence

Complex ecologies of skill include components, resources and practices that become so familiar to performers that they may pass unremarked or unnoticed. Practitioners need not be fully aware of their diverse support systems when all is going well, and many forms of social, technological or environmental 'scaffolding' may become transparent in use. But this doesn't mean that the scaffolding isn't present. It may be useful to retain a background awareness of the elements of wider ecologies of skill and to be able swiftly to refocus attention on them if required (compare Wheeler 2019). Given that expert performers typically resist the full automation of their skills and train themselves to cope effectively even in unfamiliar or challenging circumstances, their own knowledge of the full ecology may be less tacit than it first seems. Specific observation or probing can, in fact, elicit access to dynamic, task- and situation-specific embodied knowledge (McIlwain and Sutton 2015). And sometimes, events themselves may make the scaffolding visible.

Alongside invocations of skilled performance as metaphor in early 4E research, philosophers and cognitive theorists often called for more ethnographic investigation of creative collaborations in the arts and sport to produce rich, detailed accounts of skill in real-world contexts. Several of our contributors do just this: they deploy observational, participant or apprenticeship methods (see also Bicknell 2021; Downey, Dalidowicz, and Mason 2015; Hjortborg and Ravn 2020; Kimmel 2021; Pini and Sutton 2021; Pink and Morgan 2013; Rietveld and Brouwers 2017; Samudra 2008; Wacquant 2015). Others use established methods of archival research, and a range of analytic methods from philosophy, dynamical systems theories and interaction studies. Motivating these approaches, in common across many distinctive methods, is a commitment to specificity and precision, to attempting thicker, experience-near description of collaborative embodied performance in action.

The ecologies under investigation may be transformed by deliberate interventions, as when Rietveld Architecture-Art-Affordances (RAAAF)'s large-scale architectural designs create new possibilities for engagement with existing sites. In other cases, subtle features of existing scaffolding may show up and change as practitioners or researchers observe or participate, as in manipulations of bodies and machines by creative artists, or in an unexpected training routine in aikido. Other disruptions are more accidental, but reveal just as clearly the structures and dynamics of a distributed ecology. Tribble's analysis of one episode at the Globe is a case in point. It also confirms that researchers should attend to glitches, to cases in which smooth collaborative action is surprised or upset, to see how systems knit together, how groups recover – individually and jointly – from interruption and disturbance (Throop and Duranti 2015). Because skilled performers often push the envelope, as they say, expanding their repertoires to include creative responses and novel challenges, they are thereby vulnerable to various forms of breakdown, jitter or choking. Collaborative embodied skills are precarious: but differently so in each domain, each case, each circumstance. Symmetries are always being broken and redistributed in the constant emergence of new forms of order. Experts' technical aptitude is often backed by effective cognitive, emotional, interpersonal and situational strategies to support fluent repair of momentary or persisting trouble. A core component of effective training regimes – for breath-holding or muay thai, for example – will be preparing to deploy available resources, whether physiological or motivational, whether reliant on equipment or on peer support, to resist or manage severe challenge. To understand the distributed resilience of these collaborative systems, our contributors tap in to their operation at a number of scales. We turn now to a brief description of the book's contents, showing how the chapters approach these rich systems.

The book

We group our ten short, focussed chapters, all around 5,000 words, into three sections or thematic clusters, which zoom in to increasingly fine-grained features of skilful interactions. All three sections, and the book as a whole, end with short commentaries by leading skill researchers, who riff on emerging themes to evoke and provoke further connections for the cognitive humanities, and pick out broader issues about psychology, performance and the spaces between disciplines.

Part 1 addresses larger ecologies of embodied collaboration, in three case studies that cast the net wider than usual in studies of theatre, dance and architecture. Drawing on a rich archive of show reports filed after every performance, **Evelyn Tribble** describes the vital, coordinated actions of the Front-of-House teams at the Globe Theatre. Though performances take place in broad daylight, the management of audience attention and dynamics, dealing with fainting, vomit and more surprising disruptions in this unique setting remains invisible until a star performer intervenes. **Sarah Pini** examines Body Weather, a Butoh-inspired dance and movement practice which cultivates attention to the changing 'weather' of performers' bodies and performance ecologies. The slowness and stillness of dancers filmed moving across the dunes in a stormy wind evokes, in Pini's analysis, a form of agency embedded in the specifics of place. Unique settings are also central to the collaborative designs of the interdisciplinary studio RAAAF. In a three-way conversation between **Janno Martens,** and **Ronald** and **Erik Rietveld**, encompassing philosophy, art history and architecture, we learn how RAAAF projects transform existing sites, materials or practices in interventions which open up new possibilities for movement or expression. **Amy Cook**'s commentary on Part 1 highlights the disconcerting shifts of perspective these chapters require from both our senses and our theories as they cut across the layers of rich ecologies. For Cook, they open up unaccustomed timescales and forms of movement in space, inviting us to look at well-known phenomena through different lenses.

Moving from macro-ecologies to a middle or meso-level of small group and dyadic interaction, **Part 2** examines collaborative and socially scaffolded learning in handstand classes, freediving, aikido training and in an unusual human-AI musical collaboration. **Kath Bicknell** and **Kristina Brümmer** vividly describe the fragility and instability of skill acquisition as they tried to learn a tough new movement practice together. Sharing fatigue, failure and humour as they develop their own novel linguistic and physical repertoire, the two researchers show us how collaboration between peers operates in a specific subcommunity of practice. A group of budding divers learn together how to hold their breaths for extended periods underwater in **Greg Downey**'s chapter on expert apnoea. This highly individual ability, which requires divers to confront and adapt very basic physiological and neurological responses, develops gradually through four specific types of collaboration. **Susanne Ravn** reconsiders her longstanding aikido practice, explaining how subtle features of her changing bodily comportment and movement techniques directly reflect the norms and ideals of this remarkable Japanese tradition. Zooming in on one unusual practice session, Ravn describes the multilevel challenges of attuning to the energy, direction and timing of a highly experienced training partner. Collaborative musical agency is examined by **Tom Roberts** and **Joel Krueger** in considering Herndon and her artificial neural network, *Spawn*. By considering how this human musician works with, and thinks about, this non-human technology, Roberts and Krueger test and expand the factors that prompt attributions of creativity and agency. In her commentary on Part 2, **Emily Cross** highlights the embodied nature of many learning processes: noting the difficulties faced by those trying to develop new skills at a distance under lockdown, Cross considers the inventive resilience of communities of practice as they find new ways to collaborate in 'Covid-normal' conditions, and looks optimistically towards a more closely connected future.

The chapters in **Part 3** focus in tightly on the dynamics and microprocesses of interaction, examining forms of symmetry, synergy and entrainment across a range of movement practices. **Rachel Kallen, Margaret Catherine Macpherson, Lynden Miles** and **Michael Richardson** address the ecological dynamics of performance-environment systems in a study of the principles underlying social coordination across multiple contexts. Events that break and then redistribute symmetry can be identified in small-scale behavioural interactions, creative improvisation in music or sport and the dynamics of larger-scale social systems. The authors seek a generalizable framework and tools for understanding embodied collaboration. Analysing one remarkable sequence of actions by an expert muay thai fighter, Singpayak, **Sara Kim Hjortborg** unpacks the operations of embodied intelligence in a precarious, fast-changing environment. 'Sing' sets his trap and springs it, Hjortborg shows, by way of three interlocking dynamic decision-making processes in the ring, thinking on and with his feet in the midst of a brutally antagonistic interaction. **Michael Kimmel** and **Stefan Schneider** take a micro-analytic approach to intercorporeal interaction at close quarters, developing a rich qualitative language for understanding joint action from case studies in acrobatics, dance and martial arts. They identify principles of the structure and the dynamics of interpersonal coordination, and show how creative movement synergies assemble spontaneously through rich, multichannel embodied communication. From the chapters in Part 3, **Tony Chemero**'s commentary draws out a methodologically pluralist vision, mixing quantitative accounts of human interaction with qualitative approaches anchored in specifically human strategies, norms, goals or rules.

Catherine Stevens provides the first of two commentaries on themes from the book as a whole. She discusses the interacting forms of knowledge apparent in collaborative action, and the complex dynamics of control in multi-agent or hybrid ecologies in terms that bridge performance studies and cognitive theory. **Ian Maxwell**'s concluding commentary adds a focus on the sensibility of the ethnographer, immersed in the weave of a community's life, as a touchstone for assessing the volume's implications. Experiences of mediation, of dwelling between distinctive disciplines or cultures, Maxwell suggests, help us to reconstrue practices that seemed familiar as remarkable.

With Maxwell's encouragement, we want readers to explore and inhabit these interstitial spaces between disciplines, and to collaborate in the performance of interdisciplinary research that may challenge, unsettle and extend our understanding of each other. We invite you to enjoy this collection of essays, and, just maybe, to seek out other curious folk with whom to think, act and work together in studying and creating new and rich ecologies of performance.

References

Adams, F. and K. Aizawa (2001), 'The Bounds of Cognition', *Philosophical Psychology* 14 (1): 43–64.

Bicknell, K. (2021), 'Embodied Intelligence and Self-Regulation in Skilled Performance: Or, Two Anxious Moments on the Static Trapeze', *Review of Philosophy and Psychology* 12: 595–614.

Blair, R. and A. Cook (2016), *Theatre, Performance and Cognition*. London: Bloomsbury.
Cappuccio, M. L. (2019), *Handbook of Embodied Cognition and Sport Psychology*. Cambridge, MA: MIT Press.
Christensen, W., K. Bicknell, D. McIlwain and J. Sutton (2015), 'The Sense of Agency and its role in Strategic Control for Expert Mountain Bikers', *Psychology of Consciousness: Theory, Research, and Practice, 2 (3)*: 340–53.
Clark, A. (1997), *Being There: Putting Brain, Body, and World Together Again*. Cambridge, MA: MIT Press.
Clark, A. and D. Chalmers (1998), 'The Extended Mind', *Analysis 58 (1)*: 7–19.
Cowley, S. J. (2002), 'Why Brains Matter', *Language Sciences, 24 (1)*: 73–95.
De Jaegher, H., A. Peräkylä and M. Stevanovic (2016), 'The Co-Creation of Meaningful Action: Bridging Enaction and Interactional Sociology', *Philosophical Transactions of the Royal Society B: Biological Sciences 371 (1693)*: 20150378.
Downey, G., M. Dalidowicz and P. H. Mason (2015), 'Apprenticeship as Method: Embodied Learning in Ethnographic Practice', *Qualitative Research 15 (2)*: 183–200.
Dreyfus, H. L. (2007), 'The Return of the Myth of the Mental', *Inquiry 50 (4)*: 352–65.
Fitzgerald, D. and F. Callard (2015), 'Social Science and Neuroscience Beyond Interdisciplinarity: Experimental Entanglements', *Theory, Culture & Society 32 (1)*: 3–32.
Fridland, E. (2014), 'They've Lost Control: Reflections on Skill', *Synthese, 191 (12)*: 2729–50.
Geeves, A., D. J. McIlwain, J. Sutton and W. Christensen (2014), 'To Think or Not to Think: The Apparent Paradox of Expert Skill in Music Performance', *Educational Philosophy and Theory 46 (6)*: 674–91.
Hahn, T. and J. S. Jordan (2014), 'Anticipation and Embodied Knowledge: Observations of Enculturating Bodies', *Journal of Cognitive Education and Psychology 13 (2)*: 272–84.
Hart, F. E. and B. A. McConachie (Eds.) (2010), *Performance and Cognition: Theatre Studies and the Cognitive Turn*. London: Routledge.
Hjortborg, S. K. and S. Ravn (2020), 'Practising Bodily Attention, Cultivating Bodily Awareness – a Phenomenological Exploration of tai chi Practices', *Qualitative Research in Sport, Exercise and Health 12 (5)*: 683–96.
Hurley, S. L. (1998), *Consciousness in Action*. Cambridge, MA: Harvard University Press.

Hutchins, E. (1995), *Cognition in the Wild*. Cambridge, MA: MIT Press.
Hutchins, E. (2010), 'Cognitive Ecology', *Topics in Cognitive Science 2 (4)*: 705–15.
Johnson, C. M. (2015), 'The Cognitive Ecology of Dolphin Social Engagement', In D.L. Herzing and C.M. Johnson (eds), *Dolphin Communication and Cognition*, 229–56, Cambridge, MA: MIT Press.
Kemp, R. and B. McConachie (Eds.) (2018), *The Routledge Companion to Theatre, Performance and Cognitive Science*. London: Routledge.
Kimmel, M. (2021), 'The Micro-Genesis of Interpersonal Synergy: Insights from Improvised Dance Duets', *Ecological Psychology 33 (2)*: 106–45.
Lutterbie, J. (2019), *An Introduction to Theatre, Performance, and the Cognitive Sciences*. London: Bloomsbury.
McIlwain, D. and J. Sutton (2014), 'Yoga from the Mat up: How Words Alight on bodies', *Educational Philosophy and Theory 46 (6)*: 655–73.
McIlwain, D. and J. Sutton (2015), 'Methods for Measuring Breadth and Depth of Knowledge', In D. Farrow and J. Baker (eds), *The Routledge Handbook of Sport Expertise*, 221–31, London: Routledge.
Menary, R. (2006), 'Attacking the Bounds of Cognition', *Philosophical Psychology 19 (3)*: 329–44.
Michaelian, K. and J. Sutton (2013), 'Distributed Cognition and Memory Research: History and Current Directions', *Review of Philosophy and Psychology 4 (1)*: 1–24.
Montero, B. G. (2016), *Thought in Action: Expertise and the Conscious Mind*. Oxford: Oxford University Press.
Newen, A., L. De Bruin and S. Gallagher (Eds.) (2018), *The Oxford Handbook of 4E Cognition*. Oxford: Oxford University Press.
Pearlman, K., J. MacKay and J. Sutton (2018), 'Creative Editing: Svilova and Vertov's Distributed Cognition', *Apparatus: Film, Media, and digital cultures in Central and Eastern Europe 6*. URL: http://www.apparatusjournal.net/index.php/apparatus/article/view/122
Pini, S. and J. Sutton (2021), 'Transmitting *Passione*: Emio Greco and the Ballet National de Marseille', In K. Farrugia-Kriel and J. Nunes Jensen (eds), *The Oxford Handbook of Contemporary Ballet*, 595–612, Oxford: Oxford University Press.
Pink, S. and J. Morgan (2013), 'Short-Term Ethnography: Intense Routes to Knowing', *Symbolic Interaction 36 (3)*: 351–61.
Port, R. F. and T. Van Gelder (Eds.) (1995), *Mind as Motion: Explorations in the Dynamics of Cognition*. Cambridge, MA: MIT Press.
Rietveld, E. and A. A. Brouwers (2017), 'Optimal Grip on Affordances in Architectural Design Practices: An Ethnography', *Phenomenology and the Cognitive Sciences 16 (3)*: 545–64.
Samudra, J. K. (2008), 'Memory in Our Body: Thick Participation and the Translation of Kinesthetic Experience', *American Ethnologist 35 (4)*: 665–81.

Sawyer, R. K. and S. DeZutter (2009), 'Distributed Creativity: How Collective Creations Emerge from Collaboration', *Psychology of Aesthetics, Creativity, and the Arts 3 (2)*: 81–92.

Shaughnessy, N. and P. Barnard (Eds.) (2020), *Performing Psychologies: Imagination, Creativity and Dramas of the Mind*. London: Bloomsbury.

Smart, P., R. Heersmink and R. W. Clowes (2017), 'The Cognitive Ecology of the Internet', In S.J. Cowley and F. Vallée-Tourangeau (eds), *Cognition Beyond the Brain*, 251–82, Berlin: Springer.

Sutton, J. (2007), 'Batting, Habit and Memory: The Embodied Mind and the Nature of Skill', *Sport in Society 10 (5)*: 763–86.

Sutton, J. (2010), 'Exograms and Interdisciplinarity: History, the Extended Mind, and the Civilizing Process', In R. Menary (ed.), *The Extended Mind*, 189–225, Cambridge, MA: MIT Press.

Sutton, J., D. McIlwain, W. Christensen and A. Geeves (2011), 'Applying Intelligence to the Reflexes: Embodied Skills and Habits between Dreyfus and Descartes', *Journal of the British Society for Phenomenology, 42 (1)*: 78–103.

Sutton, J. and K. Bicknell (2020), 'Embodied Experience in the Cognitive Ecologies of Skilled Performance', In E. Fridland and C. Pavese (eds), *The Routledge Handbook of Philosophy of Skill and Expertise*, 194–206, London: Routledge.

Sutton, J. and N. Keene (2017), 'Cognitive History and Material Culture', In D. Gaimster, T. Hamling, and C. Richardson (eds), *The Routledge Handbook of Material Culture in Early Modern Europe*, 44–56, London: Routledge.

Throop, C. J. and A. Duranti (2015), 'Attention, Ritual Glitches, and Attentional Pull: The President and the Queen', *Phenomenology and the Cognitive Sciences 14 (4)*: 1055–82.

Toner, J., B. Montero and A. Moran (2021), *Continuous Improvement: Intertwining Mind and Body in Athletic Expertise*. Oxford: Oxford University Press.

Tribble, E. and N. Keene (2011), *Cognitive Ecologies and the History of Remembering: Religion, Education and Memory in Early Modern England*. London: Palgrave.

Tribble, E. and J. Sutton (2014), 'Interdisciplinarity and Cognitive Approaches to Performance', In N. Shaughnessy (ed.), *Affective Performance and Cognitive Science: Body, Brain, and Being*, 27–37, London: Bloomsbury.

Wacquant, L. (2015), 'For a Sociology of Flesh and Blood', *Qualitative Sociology 38 (1)*: 1–11.

Wheeler, M. (2019). 'The Reappearing Tool: Transparency, Smart Technology, and the Extended Mind', *AI & Society 34 (4)*: 857–66.

PART ONE

Complex ecologies of embodied collaboration

1

Dropping like flies: Skilled coordination and Front-of-House at Shakespeare's Globe

Evelyn B. Tribble

This volume asks us to consider the dizzying variety of ways that tightly knit collaboration works in unique places and settings. Most discussions of skilled theatrical performance naturally focus upon the artistic production itself – what are the constraints within which a company works (Tribble 2011)? How does a production get on its feet, and how is meaning co-created during the rehearsal process (McAuley 2008)? Studies of 'distributed creativity' (Sawyer and DeZutter 2009) have described creativity not as an individual property but as part of an emergent, collaborative system. This methodology has sparked further research, often taking the form of observing the 'ecological dynamics' of the rehearsal process (Clarke et al. 2013, 628; Parolin and Pellegrinelli 2020), including the social and material constraints of such activities.

As valuable as these studies are, they often stop at the rehearsal room door and do not always extend to an entire cognitive ecology of performance: the physical, material, social, affective and mental resources that sustain and maintain it. Cognitive ecologies are dynamic, changing to accommodate new circumstances: some systems will place more or less weight on internal mechanisms, on central control or on particular forms of cognitive scaffolds and social systems. The framework of cognitive ecology is derived from a range of disciplines, including anthropology (Hutchins 1995, 2014), ecological psychology (Gibson 1979); philosophical accounts of 4E cognition (embodied, embedded, enactive and extended); the rich literature on distributed cognition, as represented by Edwin Hutchins's groundbreaking study *Cognition in the Wild* (1995); and the related field of distributed creativity (Sawyer and DeZutter 2009). A cognitive ecological approach to performance does more than simply confirm the truism that theatre is a collaborative art. Theatrical performance always takes place in real time, requiring adjustments on the fly and the quick assemblage of explicit and implicit knowledge. We are familiar with the way that skilled actors and stage managers handle the unexpected and ensure a smooth flow of performance despite unexpected events, including forgotten lines, misplaced props and unruly or inattentive audiences. Despite being absolutely essential to any performance, audiences are often seen as somehow extraneous, factors to be controlled rather than integral to the ecology of performance. Indeed, modern acting training often urges actors to 'forget the audience, to act as if they are not even there' (Paulus 2006, 335). If the audience itself is often relegated to an extraneous factor, even more neglected is way that audiences are managed: the skilled coordination exhibited by Front-of-House (FOH).

Focusing on Front-of-House challenges conventional understandings of what counts as performance, what counts as thinking and how status operates in organizations. Integrating FOH into models of collaborative skilled performance expands the frame of analysis beyond the artistic product to encompass the larger 'art world'. Howard S. Becker coined this term to describe the nature of the 'joint activity' underpinning creative work (Becker 2008). Becker defines 'art worlds' broadly, encompassing both contemporary and historical theatre and dance, the relationship of publishing institutions and literature and the visual art market. Provocatively, he begins the first chapter with Anthony Trollope's coffee, brought to him by a servant every day at 5.30am when he began his daily work. Thinking about Trollope's coffee, the labour involved in producing it and the work that it enables, Becker argues, produces 'an understanding of the complexity of the cooperative networks through which art happens' (Becker 2008, 1). A full study of the 'joint action' of an art world requires that the research 'look for all the people involved ... especially the ones conventionally thought not to be very important' (Becker 2008, ix); Becker mentions stagehands, ticket takers, parking attendants and similar agents, who are often thought of as 'support', a 'residual category, designed to hold whatever the other categories do not make an easy place for' (Becker 2008, 4).

Becker does not fully explore how such 'residual' categories might be integrated into a full account of a given art world, save to suggest that all elements are essential, however taken for granted or in the background they may be. One approach to such an integration is the concept of 'collaborative emergence'. Keith Sawyer and Stacy DeZutter distinguish this from routine and predictable activity, which may be collaborative, but is not emergent. Collaborative emergence requires not just 'routines and procedures', but a 'moment to moment contingency', allowing 'something novel and appropriate to occur' (Sawyer and DeZutter 2009, 81). As an example, they cite Hutchins's (1995) account of an emergency aboard a naval vessel, which required a sudden shift from 'well established group routines' to a 'collaboratively creating a novel, improvised response' (Sawyer and DeZutter 2009, 83).

In this chapter, I argue that when studied within the framework of cognitive ecology, FOH can be seen as integral rather than peripheral to the art world of contemporary theatre. Because of its focus on thought as embodied and extended into its environment, and because of its focus on emergent practices, a cognitive ecological approach allows a view of the often-invisible resources that underpin on-the-fly coordination in real time. Typically neglected in accounts of performance, FOH might be described as the retail of theatre, the least glamorous but absolutely essential element of any theatrical event. Led by the House Manager, FOH manages the audience, times the incoming (the moment when the audience is allowed into the theatre), handles complaints, keeps order, tells the stage manager when to take the show, corrals patrons after intermission, manages expectations, oversees health and safety and responds to catastrophes large and small, routine and unusual, including fainting and illness, inattentive or disruptive school groups, entitled donors, drunken patrons, mishaps and accidents, crying babies, surly patrons, illicit cell phone use, and on and on.

FOH must coordinate seamlessly with a range of other systems and entities. Modern theatres are complex entities, comprising education, outreach, box office, concessions and gifts, fundraising, grant writing, publicity, financials and more. Each of these has its own miniature ecosystem reaching out and intersecting with others but with distinct sets of practice, modes of working and relationships to the larger system. Compared to the work of the company and the cultural prestige of the theatre itself, FOH has relatively low status, yet its work demands exquisite coordination in real time, expert knowledge of the theatre, tact, patience, firmness and a tolerance for bodily fluids.

I focus upon a particularly rich and complex case study: FOH at Shakespeare's Globe, based upon the datasets of show reports written and filed after every performance, from the opening of the theatre in 1997 to 2016.[1] These records document the unfolding and emerging efforts to manage a new kind of audience for a new kind of theatre. The intent of the planners was to recreate as far as possible the original staging conditions in which Shakespeare's plays were performed, and thereby to explore or rediscover Shakespeare's stagecraft and dramaturgy. The Globe is constructed as an open-air theatre, with a large Yard for standing audience members, and benches arrayed in gallery structure on three sides. The thrust stage has substantial pillars holding up a roof that protects the actors from rain, but the Yard itself is open to the elements. Audiences and actors share the same ambient lighting, and audiences are fully visible both to the performers and to one another.

The design of the theatre means that the audience can play a much larger role in this environment than in most contemporary theatres. It is often said that the oxygen of the Globe is its audience. Actor/audience engagement has been one of the most distinctive features of the theatre. 'The actors as well as the audience have had to relearn their roles, and new conventions of communication with the audience have been devised' (Carson and Karim-Cooper 2008, 180). The director Tim Carroll described the Globe audience not as passive, but as 'the most versatile scene partner in the world' (Carroll 2008, 40). Globe directors train actors to work with the audience – to address their remarks not just to the groundlings in the Yard but also to spectators across the theatre, including the upper galleries. This approach can create a raucous atmosphere that has often been subject to critique. Indeed, initial responses to the new theatre were sceptical, likening it to 'a cultural theme park, a cross between Disneyland and a National Trust property' (Carson 2008, 115). Critics focused upon the often-rambunctious audience members, comparing their loud and boisterous reactions to the atmosphere at pantomimes and suggesting that the audiences are encouraged to engage in acts of spurious 'simulation' of Elizabethan audiences (Silverstone 2005, 42).

In contrast, Penelope Woods has argued that these audience members can be considered skilful subjects rather than unwitting pawns in the marketing of the theatre. In this context,

> skill may not be simply an isolated or individual capacity, the sole preserve of the performer in the theatre, but might instead be seen as distributed across the 'system' of theatre. The audience is essential for the realisation of performance. Where the audience is visible and participatory they are more substantively intrinsic to this 'system'; their contribution has greater consequence and should be seen as potentially 'skillful'.
>
> (Woods 2015, 100)

This account of audience skill and agency is a welcome corrective to reception accounts of the audience. However, while the space between actor and audience may be a contested terrain, it is not simply bidirectional. Hidden from this equation are the collaborative embodied skills needed to manage and mediate that relationship. FOH is largely responsible for thinking and responding in real time to the fluid and dynamic elements of the space.

Sometimes FOH is seen simply as a form of managerialism. Silverstone argues that 'While encouraging audience participation, the Globe seems anxious that its spectators behave in an "appropriate" manner. To this end the Globe works to discipline the spectators. Tellingly, most of this discipline is meted out against the groundlings, perhaps as a late capitalist comment on their cheaper "seats"' (2005, 43). FOH is here positioned as mere disciplinary functionaries of the larger Globe enterprise. But this account neglects the skill and agency of FOH and its crucial role in the creative collaborative project of the theatre. The actors work to command the attention of the audience; FOH works to manage that attention. Certainly, FOH must work within the framework of top-down decisions, including branding, marketing, artistic direction, casting and so on. We might think of these as vertical systems, represented explicitly by organizational charts and operating manuals and implicitly by status hierarchies. FOH copes with the downstream here-and-now effects of decisions made by these agents. The reception of these vertical decisions occurs horizontally on the ground and in real time through myriad small interactions between FOH and patrons. Thus the work of FOH is, literally, displayed on the ground, and attention to this dimension reveals the interplay between standardized and improvised on-the-fly work, the particular skilled practices elicited by the task demands and the friction that sometimes occurs at the interstices or contact zones between groups.

One moment of such friction sheds light on tensions between top-down status-driven hierarchies and such horizontal on-the-ground systems. Probably the most common problem managed by FOH is illness of various kinds, especially fainting and vomiting. The theatre is often packed, shows are put on in mid-afternoon on hot summer days, standing can be physically demanding and patrons are often unwell. On one warm day in June the show report noted that 'as expected, audience members were dropping like flies' (*Twelfth Night* 16 June 2002). Illness often has a ripple effect: 'Tonight's audience is clearly going for the record on how many people can faint and vomit during one performance. We're up to about 15 at last count. No time to write an amusing report tonight, as I need to go and wash the vomit off my shoes now. Nice' (*Cymbeline* 26 July 2001). Very few performances pass without at least one incidence of illness, and the show reports remark on the anomaly of illness-free events. FOH develop protocols to assist patrons, helping them out of the theatre, providing them with a place to recover, even laundering their clothes for them. In serious medical situations, they summon ambulances.

In most situations, these systems work so well that most of the audience is unaware of the disruption. The relatively fluid space of the Globe and the ambient light and noise means that such moments are much less obvious than they would be in a darkened proscenium theatre. When these systems are allowed to work, they are relatively seamless. But one series of incidents reveals that top-down interference can collapse the system. On 7 July 2001, during a production of *Cymbeline*, 'a lady fainted in the Middle Gallery. [Artistic director and lead actor] Mark Rylance stopped the show while a steward and front of house brought her out' (7 July 2001). The next day, this intervention was repeated:

> 10 minutes before interval a lady in the Middle Gallery fainted and hit her head. FOH and security were there almost immediately. Mr. Rylance once again stopped the show causing unnecessary confusion amongst the people around her. One lady decided to call an ambulance from her mobile but FOH had a situation in hand and asked her to cancel the call. She [the fainter] was brought out of the theatre in into the first aid room by front of house and security, but by that time Mr. Rylance called for a 10-minute break. That meant that the piazza [lobby] was now full, and neither the stewards nor Milburn's [the caterer] were ready. It also meant we had a full piazza while waiting for an ambulance. The impromptu break was then made into the interval at our request as we felt this would be less disruptive for the rest of the audience. The lady was treated by security and front of house in the first aid room until the paramedic arrived ... The audience were slow to go in after the interval and the play resumed. People being taken ill in the theatre is a common occurrence at the Globe which front of house deal with on a regular basis. We have procedures designed to lessen the disruption and embarrassment for the party and to get professional help to the person as quickly as possible in accordance with our health and safety regulations. The decisions today made it difficult to do to this.
>
> (*Cymbeline* 8 July 2001)

This incident shows that horizontal ecologies can be completely disrupted by vertical interventions. As artistic director and lead actor, Rylance wields enormous authority; stopping the show and directing a 10-minute break threw the system completely out of whack, escalating the incident and creating multiple points of strain, including the disruption of catering and preparations for admission. The audience member's attempt to ring an ambulance shows the results of ambiguous lines of authority, and calling an unscheduled break meant that the piazza/lobby area was teeming with patrons when the ambulance did arrive. The admonition at the conclusion of the report is unusually direct in its claims for the expertise of FOH, even while being couched in passive voice that does not re-identify the high-status agent responsible for the 'decisions'.

Becker argues that it is possible to assess the importance of 'support' activities in art worlds by imagining their effects of their absence. This incident stages just such a moment – when the functions of FOH are usurped, their collaborative skill becomes visible in its very absence. This moment also sheds light on the nature of the collaborative expertise of Globe FOH as they negotiate the boundaries of the actor/audience interaction; actively manage the space; and negotiate the downstream effects of decisions made upstream.

FOH at the Globe must manage a space that breaks with many of the contemporary conventions of professional theatrical space. The space at the Globe is markedly different from a traditional indoor proscenium theatre; in fact, its dynamics in some ways bear more resemblances to a sporting event than to a traditional theatrical event. In contrast, in most modern indoor theatres and cinemas the space and seating arrangements themselves are designed to manage the audience and to separate audience members both from one another and from the performers. In such spaces, seats are clearly marked and delimited by armrests. The system has been designed so that some of the work of settling audiences is offloaded onto the space itself. Of course, FOH in these settings must still manage routine problems such as late arrivals and early departures, rustling of candy wrappers, cell phone use and the like. However, in modern indoor theatres, the audience is relatively stationary and contained once it is finally seated, especially when the lights are dimmed for performance. One could say from the opposite perspective that irritations from noise and other breaches of etiquette are amplified in a context in which the audience is to remain seated, fixed and quiet in a darkened space. At the Globe, by contrast, the background hum of noise and commotion means that audience members have to be particularly noisy or unruly to draw attention – a challenge that some audience members nevertheless rise to.

Because the Globe does not partially offload audience management onto the design features, the space has to be actively monitored and controlled. The freewheeling nature of the theatrical space – the ability of standing patrons to move relatively freely and to claim their own space – is crucial to the dynamic of Shakespeare's Globe. Canny 'groundlings', those who have paid five pounds for their standing tickets, line up outside the theatre early and make a beeline for the edge of the stage, where they are permitted to lean their arms. There is often a kind of background hubbub of patrons entering and leaving, jockeying for better positions and moving out of the way of actors, who sometimes use the Yard for entrances. Those who pay the much higher prices for seats find long rows of wooden benches with no backs or armrests. The seat numbers are marked but not physically delimited, which can cause problems when audience members attempt to spread out or squeeze standing friends in, so even seated space has to be constantly managed and negotiated. Moreover, the Globe also must adhere to modern health and safety regulations, including capping audience numbers and ensuring that entry ways and exits are not blocked. These dynamics have to be constantly monitored and negotiated, through a system of trained volunteer stewards reporting to the FOH staff and the house manager. The Globe is unusual among theatres in that spectators are actually able to claim small bits of the stage as their own, rather than being separated from it. Audience members sometimes nick small property items and in turn are sometimes nicked by the properties wielded by actors: 'One wee punter was trying to steel [sic] Macbeth's crown on the outgoing, but I said I'd have to have it back thank you very much' (*Macbeth* 7 June 2001); 'Lady was hit in the face by a flying fish. FOH apologised' (*Comedy of Errors* 10 June 1999).

Actors and directors encourage the freewheeling interaction with the audience, yet also seek help from FOH to quiet overly rambunctious audience members who think that rude loud behaviour is somehow required. Some patrons believe that they are expected to play the role of rowdy playgoers in Shakespeare's time, as when some 'naughty groundlings were chucking fruit at Malvolio in Act 2.' When asked to stop, 'cheeky lady replied, "Oh, I thought it was all part of the thing"' (*Twelfth Night* 18 May 2002). This sense of the 'thing' as a form of role-playing was especially apparent when a large group of Tudor cos-players, dressed in period costume, took up three seats each to accommodate their costumes and refused to 'take their feathery headdresses off' (*Romeo and Juliet* 24 September 2004).

Sometimes these incidents are handled by the actors by bending the frame of the experience. The fluid interaction between actor and audience that marks the Globe experience can be recruited to manage difficult audience members; at these points a delicate balancing act determines whether hecklers or fainters are managed by the company and thus integrated into the show or left to FOH to discreetly handle. Sometimes these interactions go very wrong, as when an actor told children in a school group to 'shut the fuck up', which prompted numerous complaints from the audience, some of whom walked out (*Macbeth* 7 June 2001). At other times, the actors manage to handle such incidents without breaking flow or frame: 'Lady heckler at the top of the show/drunk & v. loud. Dealt with very ably by company members. FOH not asked to intervene so let it go with the flow' (Middleton, *Chaste Maid in Cheapside* 2 September 1997).

While actors can 'go with the flow' and envelope audience management into their performance, FOH intervention requires a wilier approach. Most problems arise from very pragmatic considerations: it can be very hard to monitor illicit photography, to keep people standing up throughout the performance, to moderate disruptive and inattentive school groups, and to manage an audience often in near-constant motion. Because the Yard is flat, children and smaller patrons can find it very difficult to see, so stewards constantly remind patrons not to stand on the raked aisles (a fire hazard) or to raise small children on their parent's shoulders. Most of these are resolved routinely, but the stewards and FOH frequently endure hostility and abuse: 'Bad tempered audience. Not helped by the freaky weather. People refusing to move off stairs and doorways and front of house apologize to people sitting around door to entrance for any disturbance. Horrible day' (*Comedy of Errors* 31 July 1999). The show reports describe many such examples of disrespect accorded to FOH as they attempted to police bad behaviour, both on the part of school groups and entitled adults. Visible marks of authority differentiating them from the apron-clad stewards provide some means of establishing credibility and demonstrating that the incident has escalated. One show report noted that 'the fact you have a walkie-talkie in your hand works wonders – don't mess with us' (*Cymbeline* 30 June 2001). The following year, their authority-markers were upgraded: 'several photographers were brought to our attention by SM [stage manager]. Thanks to our new headsets, we looked like a crack Army unit swooping for the kill. One woman in question looked rather alarmed and she didn't do it again' (*Twelfth Night* 21 June 2002).

Another element of status that operates in Shakespeare's Globe is that of Shakespeare himself. Front-of-House must manage competing expectations around the nature of the experience, expectations that are in part generated by the branding of the theatre as a serious theatre, a site for research and education and a place for heritage tourism. Many of the audience view the Globe as a tourist site and do not understand why they are not able to take photographs at will: 'Some passing sightseers were very surprised that they could not just pop in to take photos and have a look around the theatre we explained that there was actually performance taking place this evening, but they didn't seem to think this is relevant apparently we're just mean' (*Cymbeline* 14 July 2001).

One of the most contentious of these interactions occurs around the 'ownership' of Shakespeare, especially as the Globe moved away from the so-called original practices of its early years under the artistic direction of Mark Rylance. The 2001 season marked a shift in design and approach to staging; as Lois Potter noted in her 2001 review, 'all the directors seem to have been encouraged to do whatever they liked to conceal the fact that they were performing on a reconstructed Renaissance stage' (Potter 2002, 95). Tim Carroll's *Macbeth* was produced in modern dress, which upset numerous patrons who felt cheated of the true Globe experience and complained that Shakespeare was somehow violated. As one of the first modernized productions performed at the Globe, the 2001 *Macbeth* prompted numerous complaints about its lack of 'authenticity'. Early in the run patrons walked out: 'complaining bitterly about the "modernness" of the production [there was] ... another complaint after 15 minutes, this time a woman who could not be pacified ... she exploded saying she had brought her two children to see a classic Shakespeare in a classic setting and didn't expect to see a show best suited to the London fringe' (*Macbeth* 31 May 2001). The same run saw FOH cope with 'one completely hysterical family whose children sobbed throughout the performance because they expected doublet and hose' (*Macbeth* 20 June 2001) and a 'loudly spoken Yorkshire woman' who hated everything and left the show 'shouting "Sacrilege" at the top of her voice' (*Macbeth* 7 August 2001). FOH bears the brunt of such disappointed expectations, mediating between artistic and marketing decisions and the assumptions of the audiences: 'One person came storming out a few minutes into the performance saying it wasn't Shakespeare. FOH explained that it was' (*Macbeth* 31 July 2001). These many moments of improvised explanations, accommodations, resolutions and pacifications represent ubiquitous yet often unnoticed moments of the online thinking and creative improvisation that characterizes FOH at the Globe.

The expertise and on-the-fly thinking of cognitive ecologies such as FOH in any performance context is often invisible and therefore subject to erasure. The bumpiness and unsettled nature of the space at the Globe makes it an apposite site to study the skill and coordination of FOH, which in other, more conventional, settings remains more persistently invisible. The parameters of these interactions are set by the goals of the larger institution, one that places great value on the 'buzziness' that characterizes the Globe theatre. And unlike the performative and presentational nature of the company, FOH must do its work (invisibly), supplying the oxygen to the audience that the theatre needs to breathe. A cognitive ecological approach permits attention not just to the most prestigious and visible elements of the system, but the on-the-ground mechanisms that underpin the larger enterprise. If we return to the account of 'collaborative emergence' proposed by Sawyer and DeZutter (2009), it is evident that FOH work is anything but routine and predictable. As in any complex system, routines and procedures underpin its activity – indeed, routines and procedures are often themselves creative responses to unexpected events – but the very essence of FOH at the Globe is managing the 'moment-to-moment contingency' (Sawyer and DeZutter 2009, 81) of the theatrical event. Because actors, audience and FOH are all in a sense performing in a shared space, equally visible to one another, the system exhibits 'the moment-to-moment dependency on the behaviours of other individuals' that Sawyer and DeZutter describe as characteristic of collaborative emergence (2009, 84).

One show report after a particularly hectic session recorded a patron remarking to FOH, 'You've got a funny job, haven't you?' (*Macbeth* 10 August 2001). Is FOH at the Globe 'funnier' than such roles in more conventional theatrical spaces? Perhaps work in such contexts may indeed come closer to routinized, scripted encounters than an embedded and extended cognitive ecology such as that I've described in this chapter. But I suspect not, although this hypothesis would have to be tested by close observation. Possible approaches to such further studies could include cognitive ethnography or the methodology of interaction analysis that Sawyer and DeZutter model. Perhaps attention to other such 'funny jobs' might reveal that such skilled coordination on (and around) the ground(lings) is not incidental to creative projects either.[2]

Notes

1 These Show Reports are housed in the Shakespeare's Globe Archive maintained by Adam Matthew: https://www.shakespearesglobearchive.amdigital.co.uk/. The show reports are at GB 3316 SGT/THTR/SR. Each show report is cited parenthetically by play and date. All reports were accessed in January 2021.
2 My thanks to John Sutton and Kath Bicknell for their patient and insightful editing throughout this project. I am also grateful to Amy Cook and Kate Rossmanith for their generous and helpful readers' reports, as well as to Rhonda Blair for reading drafts and Will Tosh of Shakespeare's Globe for his encouragement. My fellow contributors to the volume provided invaluable feedback in our work-in-progress session. I would also like to thank the Early Modern Studies Institute at USC for inviting me to present this research at their Huntington seminar.

References

Becker, H. S. (2008), *Art Worlds: Updated and Expanded*. Berkeley: University of California Press.
Carroll, T. (2008), 'Practising Behaviour to His Own Shadow,' in C. Carson and F. Karim-Cooper (eds), *Shakespeare's Globe: A Theatrical Experiment*, 37–44, Cambridge: Cambridge University Press.
Carson, C. and F. Karim-Cooper (Eds.) (2008), *Shakespeare's Globe: A Theatrical Experiment*, Cambridge: Cambridge University Press.
Carson, C. (2008), 'Democratising the Audience?' In C. Carson and F. Karim-Cooper (eds), *Shakespeare's Globe: A Theatrical Experiment*, 115–26, Cambridge: Cambridge University Press.
Clarke, E., M. Doffman and L. Lim (2013), 'Distributed Creativity and Ecological Dynamics: A Case Study of Liza Lim's "Tongue of the Invisible"', *Music and Letters* 94 (4): 628–63.
Hutchins, E. (1995), *Cognition in the Wild*, Cambridge, MA: MIT Press.
Hutchins, E. (2014), 'The Cultural Ecosystem of Human Cognition', *Philosophical psychology* 27 (1): 34–49.

Gibson, J. J. (1979), *The Ecological Approach to Visual Perception.* Boston: Houghton Mifflin.

McAuley, G. (2008), 'Not Magic but Work: Rehearsal and the Production of Meaning,' *Theatre Research International* 33 (3): 276–88.

Parolin, L. L. and C. Pellegrinelli (2020), 'Unpacking Distributed Creativity: Analysing Sociomaterial Practices in theatre Artwork', *Culture & Psychology* 26 (3): 434–53.

Paulus, D. (2006), 'It's all about the Audience', *Contemporary Theatre Review* 16 (3): 334–47.

Potter, L. (2002), 'The 2001 Globe Season: Celts and Greenery,' *Shakespeare Quarterly* 53 (1): 95–105.

Sawyer, R. K. and S. DeZutter (2009), 'Distributed Creativity: How Collective Creations Emerge from Collaboration', *Psychology of Aesthetics, Creativity, and the Arts*, 3 (2): 81–92.

Silverstone, C. (2005), 'Shakespeare Live: Reproducing Shakespeare at the "new" Globe Theatre', *Textual Practice* 19 (1): 31–50.

Tribble, E. (2011), *Cognition in the Globe: Attention and Memory in Shakespeare's Theatre.* London: Palgrave.

Woods, P. (2015), 'Skilful Spectatorship? Doing (or Being) Audience at Shakespeare's Globe Theatre', *Shakespeare Studies* 43: 99–113.

2

On the edge of undoing: Ecologies of agency in Body Weather

Sarah Pini

Performing agency

As an interdisciplinary researcher and dance artist, I often wonder how evocative movements and meaningful choreographies develop, and what makes a suggestive performance. What inspires moving artworks? Which are the elements involved in such creative processes and what role do they play in shaping the performance? What kinds of agencies are played out, or to put it shortly, what 'moves' us? In this chapter I provide an account of how radical forms of agency are enacted in – and through – specific dance and performance practices.

Artistic and performative practices hold the potential to shed light on fundamental roles of the body in cognition, offering original perspectives that contribute to questioning and reframing established notions of subjectivity and different forms of agency.

The field of performing arts has an extensive tradition in developing, addressing and accounting for multiple forms of agency that relate not only to the individual, but also to others (Crossley 1995; Jamieson 2020; Lucie 2020; Pini and Maguire-Rosier 2021; Wright 2011). Through the development of different methodologies for nurturing and attending to different sensory experiences and action–perception modes, performance practices offer 'tools and imaginaries that can complement other critical approaches to emergent forms of agency' (Jamieson 2020, 8).

Recently, in dance and performance studies there has been increased attention on analyses of performance practices through 4E cognition approaches (Blair and Cook 2016; Christensen et al. 2015; Pini and Deans 2021; Sutton and Bicknell 2020). Of particular relevance to my own study, the notion of agency has been addressed by 4E theorists in the context of improvisational dance practices (Bresnahan 2014; Merritt 2015; Ravn 2020; Ravn 2021). In this context, agency is understood as 'the control and intention the dance performer has to move in a certain way' (Bresnahan 2014, 86). Drawing on the work of Andy Clark (2008), dance philosopher Aili Bresnahan argues that 'agency is limited not just by historical, political, and social constraints, but that it is also limited and enhanced by our trained and naturally developed abilities to engage with the world on the fly, thinking-while-doing' (Bresnahan 2014, 87).

Developing an enactive perspective, Buhrmann and Di Paolo stress how our 'sense of agency presents itself phenomenologically as a *heterogenous* collection of different ways or aspects of feeling in control that depends on context, the task, and the person's history and capacities' (Buhrmann and Di Paolo 2017, 228 – emphasis in original). Similarly, addressing the subjective and felt sense of agency in improvisational dance, phenomenologist Susanne Ravn stresses the relevance of both the cultural and social aspects in which a dance form is practised. Ravn agrees with Bresnahan that dance training is 'to be understood as a specific kind of agentive training – equipping the dancer to move from a certain repertoire of possibilities' (Ravn 2020, 79). Ravn's (2020) analysis of agency builds on the phenomenological experience of solo improvisation by international dancer Kitt Johnson, while Bresnahan's (2014) account of improvisational artistry refers to a single dancer performing for an audience. Here, I focus instead on the accounts of several dancers of Body Weather (BW), a radical movement practice developed by Japanese choreographer Min Tanaka and members of the Maijuku performance group in the 1970s and 1980s.

In this chapter I present the case of BW to develop the notion of an 'ecological' sense of agency and show how it emerges through the practice of this dance form. I begin with an account of the historical and cultural context of this dance methodology. I then address the case of the production of *AURA NOX ANIMA*, a BW-inspired short dance film directed by Sydney-based visual artist and BW practitioner Lux Eterna, filmed on the sandy dunes of Anna Bay, New South Wales, Australia, in 2016.[1]

By addressing some salient principles and training features which characterize the practice of BW, as observed and practised during my fieldwork, I discuss how agency is understood and experienced in this context of practice. Based on numerous conversations with BW performers and members of Australian Body Weather dance company De Quincey Co in which they describe their lived experience of dancing 'stillness' in the dunes, I emphasize the way this dance methodology cultivates a somatic attention that expands beyond the body of the performer, stretching out to the broader environment (Pini 2020; Pini and Deans 2021). I argue that agency in BW emerges as a capacity for sensing and 'undoing' rather than 'controlling' the body, developed through attention to – and in relation to – the surrounding environment.

I conclude by emphasizing the relevance of environment-culture-context situatedness in the study of embodied cognition and creative artistic experience. By offering an account of the ways BW performers experience agency through practising stillness and slow movement in relation to the surrounding landscape, I suggest an ecological notion of agency is a key concept in and for the study of embodied cognition in performance.

Methods: Situating the body

This chapter is based on my direct engagement with BW, as practised during a year-long fieldwork period in New South Wales, Australia, in 2016 and 2017. I adopted a cognitive ecological framework and a phenomenological and ethnographic approach (Downey 2010; Downey et al. 2015; Leach and Stevens 2020; Pini et al. 2016; Pini and Pini 2019; Pini 2020; Pini and Deans 2021; Pini and Sutton 2021; Ravn 2009; Ravn and Hansen 2013; Tribble and Sutton 2011). This methodological approach included somatic and kinaesthetic participant observation and the analysis of the accounts of BW performers encountered during my apprenticeship.

Cognitive ecological theory provides a framework to tackle 'the relationships of the social and the material to cognitive processes that take place inside individual human actors' (Leach and Stevens 2020, 98). Cognitive anthropologist Edwin Hutchins has observed how through the cognitive ecology framework, 'perception, action, and thought will be understood to be inextricably integrated, each with the others. Human cognitive activity will increasingly be seen to be profoundly situated, social, embodied, and richly multimodal' (Hutchins 2010, 712). In this view, performance practices provide (and evidence) ways of knowing and constructing meaning that are shared among the members of a group.

Hutchins's cognitive ecology is an apt framework for understanding creative processes as they emerge, are shaped and distributed across members of a given group. Adopting a cognitive ecological approach to the study of agency in specific performance ecologies requires delving into the intricate matrix of interactions and interconnections – of perception, movements, action, senses, emotions and thought – that shape the relationship between particular social beings and complex environments. I focus on relevant aspects of the BW dance methodology as explored during several classes, workshops and dance improvisation events organized by members of De Quincey Co.[2] I also draw on several interviews with members of the BW performance community. To prompt discussion on relevant aspects of BW practice, during these interviews I often played a video performance displaying one of the interviewees' works and watched it with them. This encouraged them to reflect about their artistic approach and creative work and the ways both are inscribed into this dance form. To contextualize the practice of BW and how an ecological form of agency arises through this dance practice, the next section provides an overview of the historical and cultural context in which this movement methodology has developed.

Body Weather

BW is a radical and anti-hierarchical movement practice stemming from the Butoh tradition, an avant-garde dance form developed in Japan in the 1960s by dancers who followed the artistic direction of

Tatsumi Hijikata and Kazuo Ohno.[3] Butoh is a dance form fostering an image-based approach to movement creation. A salient aspect of this form consists in 'an emphasis on the transformation of the dancer into something else, an intense physicality that may result in explosions of movement across the stage or a strictly contained tension beneath the surface of the skin, and a focus on themes such as death, marginality, and nature' (Candelario 2019, 12).

BW arises from the Butoh tradition and was developed by Japanese choreographer Min Tanaka and his Maijuku performance group through the 1970s and 1980s.[4] Tanaka established his dance group Maijuku in 1985 and further developed his own training methodology, which only later became known as Body Weather. Described as 'a semi-intuitive technique' (Marshall 2006, 55), BW is a movement philosophy for creating dance. It involves performing and training the dancing body in relation to the broader environment. Fuller describes BW not as movement training but as research, as a way of acquiring information and stimulation from the physical environment (Fuller 2014, 2018).

According to Marshall, 'Tanaka was one of the first *butoh* artists to come to Australia, his Mai-Juku company performing at the Sydney Biennale in 1982' (2006, 56 – emphasis in original). Through such international exchanges, Butoh came to influence the Australian contemporary dance scene, with artists like Tess de Quincey going to Japan to train and perform with him. De Quincey joined Min Tanaka and his Maijuku performance group for six years, from 1985 to 1991. She introduced BW in Australia in 1989, presenting dance-performances and site-specific works in a range of different environments and contexts, including immersive research workshops in metropolitan and outback areas. She established De Quincey Co formally in 2000 as Australia's leading Body Weather company.

BW not only blends contemporary dance training with Asian and Western philosophy and training practices, but also melds the performing arts disciplines of dance and theatre with visual arts, film and music. According to Ian Maxwell and de Quincey, BW emerges exactly at the intersection between different genres and art forms, 'taking advantage of the ways in which different presentational genres – from gallery installations and black box theatres, to site-specific works of shifting scales, from industrial environments to desert riverbeds – contextualize and determine perception and reception' (de Quincey and Maxwell 2019, 167). To focus on one BW performance process in more detail, in the following section I narrow in on a specific case study: the BW-inspired short dance film *AURA NOX ANIMA*. I address the creative and relational aspects underlying its production, and how different forms of agency emerge and are shaped by specific features of BW training.

AURA NOX ANIMA (2016)

The first time I saw the short dance film *AURA NOX ANIMA* I was struck by the contrast created by the black and white photography, and the austere yet natural scene (Figure 1). On that occasion, the film director, Lux Eterna, had invited me to the screening of her work at 107 Projects in Redfern, Sydney.

AURA NOX ANIMA features BW dancers and Sydney-based performers, Angela French, Jessa Holman, Lauren Lloyd Williams, Kirsten Packham and Kathryn Puie. The film captures the slow passage of time: it is a long durational performance where the dancers engage in an exploration of agency in relation to the natural elements of the sand, the wind and the vast emptiness of the dunes.

FIGURE 1 *A snapshot from HD digital video,* AURA NOX ANIMA. © *Lux Eterna.*

Filmed during a stormy day, the bodies of the dancers appear as moved by the wind, slowly drawing a minimal choreography across the dunes. Through the practice of extended stillness, the film offers an embodied reflection on the themes of death and decay, inviting the audience to pause and dwell on the cyclical nature of life. The film calls for an acknowledgement of our existential precariousness in face of the transience of time.

During my BW training in Sydney in 2016 at the ReadyMade Works space in Ultimo, I met several BW artists and practitioners. During a conversation with Lux Eterna and Kirsten Packham, I played *AURA NOX ANIMA* and we watched it together. While the film unfolded, I asked the artists about their creation of the movement sequences and the filming process. Lux Eterna began by stressing the relevance of having previously shared the practice of BW with the group of dancers. She observed how having this shared 'language' facilitated the communication of her specific artistic intent during the shooting, as well as significantly influencing the aesthetic composition of the film. She stated, 'I feel really lucky that I have been able to work with Kirsten and the others, it was really amazing that I met [them] through BW so there was this unspoken language between us that I did not really have to direct.' Lux Eterna further commented on the fact that despite the importance of giving the performers precise instructions, because she trained in BW with them for several years prior to this project, she was able to connect with them in a way that had an impact on the entirety of the creative process and production of the film. In her words:

> [T]here is the need to be clear, to have guidelines, the directions need to be super clear, but they [dancers] just went into it [elements given] and anything that came up was exactly what I wanted or desired or enjoyed or really appreciate seeing. Although I was distant from them and behind the camera, I still felt very connected. I think it is an ongoing dialogue and it's more energetic, and I guess having been training together for almost three years – developing sensitivity to the elements – on the filming day we had the inclement weather, and we just did it, with the storm, with everything, and they just went out and that was really that practice of being in and of the environment that changes everything, so it is opening up and surrendering to what is rather than what could be.

Lux Eterna's account of the film's creative process echoes Hallam and Ingold's notion of agency manifested in creative and improvisation practices (2007). According to Hallam and Ingold, creativity 'is not distributed among all the individuals of a society as an agency that each is supposed to possess a priori ... but rather lies in the dynamic potential of an entire field of relationships to bring forth the persons situated in it' (Hallam and Ingold 2007, 7). Similarly, by addressing the production of *AURA NOX ANIMA*, in relation to salient aspects of this training method, the following sections highlight how BW dancers experience agency not as a feeling of being in 'control', but as a distributed process, as a 'field of relationships' arising from the interconnection of the performers' bodies and the landscape.

Bisoku and 'weathering': Ecological forms of agency in BW

While we were watching the film, captured as I was by the dancers moving gradually across the dunes, Kirsten interrupted this mesmerizing moment by reiterating the important role that having trained together in BW played for the creation of the film. Referring to the moment around minute five, when the dancers descend towards the sand from a standing position, Kirsten explained how even if the dancers were to perform a simple action such as slowly moving towards the ground, having trained together in BW facilitated the expression of the quality of (slow) movement that was sought by the director. This particular aesthetic quality of the performers' movements is attained in BW through the practice of slow movement, also referred as *bisoku*. The element of *bisoku* consists of training practitioners' capacity for moving at an extremely slow pace (see also Fuller 2014, 199). According to Fuller (2017), Tanaka's work with *bisoku* enables the manipulation of time in the form of slow movement (Fuller 2017, 82). During an interview with de Quincey recorded in Sydney, in May 2016, she explained *bisoku* further:

> Anything that is faster and impressive carries us through, we can go along with it but there is also that impressive aspect that

'maybe I can't do that', so there is that distance [...] whereas when you stop in space and just make a connection through stillness [...] you open a space that's big for us all, and we all participate in that ...

The practice of *bisoku* and stillness cultivates a bodily attention to the subtle changes inside and outside the bodies of the performers. This practice characterizes the entirety of *AURA NOX ANIMA*: the dancers slowly walking across the dunes, rolling on the sand, moving through and moved by the stormy wind, tuning to the natural landscape in a way that almost feels like their bodies are held by space, shaped by it.

In BW, audience, performers and environment are immersed in an atmosphere that is always changing, responding to continuously new internal and external stimuli. Given the prominent role that prompting a perceptual attentiveness to the environment plays in BW, I asked Lux about her choice for the location of the film, and how being immersed in such an environment influenced her work. Lux framed her creative process as 'weathering', which she later explained as training for deep sensitivity to the array of bodily fluctuations in relation to the environment:

Weathering to me is also the life journey. The weathering can inform marks, can inform movements, can inform a richness that gets deeper over time. But in the *AURA NOX ANIMA*, it was probably more connected to literally the weather and how it moves us. How it creates the impulse within us to move. Also, the body weather [sic], as the name suggests, is about the weather of the body. Every day that you wake up, it is its own microclimate, it is its own microcosm that's really specific and you have to learn to become attuned to what the weather of your body is in any given moment, in any particular environment. It's that sensitivity training.
(Lux Eterna cited in Dance Cinema 2019)

According to Lux Eterna, the idea of 'weathering' in a BW performative context is configured as a cultivation of fine tuning of the senses. Similarly, de Quincey and Maxwell emphasize how 'the deep energy of our bodies is embedded in space, shaped by time, the environment, the specifics of place'. In the context of BW, 'the focus on an energetic exchange between bodies dissolves the logics of inside and outside, self and environment' (de Quincey and Maxwell 2019, 167). My interviewees often stressed how this practice cultivates a heightened awareness to the inner sensations of the body as well as its surroundings. This includes an openness and attention to both perceiving and responding to these two elements. For example, in relation to the practice of *bisoku*, this openness of the senses, this enhanced attention is cultivated through 'working outside the normal scales and speeds of human body, of habit, and it's like examining of habits or becoming aware of habits, and then, trying to let go' (Kirsten Packham cited in Pini and Deans 2021). As I have discussed elsewhere, it is in the broadening of an intentional focus to the surroundings, and in the alteration of the experience of time what allows for the cultivation – and transformation – of the ways participants can be-with what arises (Pini and Deans 2021).

Performance scholars and feminist philosophers have promoted a shift in perspective that sees the world as an active subject, where 'acknowledging the agency of the world in knowledge makes room for some unsettling possibilities' (Haraway 1988, 593). De Quincey and Maxwell claim that the body of the performer in BW is reconfigured through an active and continual exchange with its surroundings. They emphasize how in this context of practice the act of performance in its exchange with the audience 'starts with the body as an environment reflecting and in dialogue with a greater environment' (De Quincey and Maxwell 2019, 166). Likewise, my interlocutors stressed how fostering an attention towards the ways the weather moves us shifts the focus from an egocentric perspective towards an ecological attunement. I argue that the experience of agency in BW performance, rather than as a feeling of being actively in control of the body, emerges as a sense of being attuned to 'the weather of your body' in relation to 'any particular environment'. By tuning in to the subtle sensations of the body, the practice of BW nurtures an awareness that departs from the body and expands beyond the flesh and bones of the dancers. As my interlocutors emphasized, 'I think even just being out in the dunes in a meditative state was part of that Body Weather experience of just allowing my body to allow the elements to move me' (Lux Eterna cited in Dance Cinema 2019). By focusing on the practice of *bisoku*, or moving slowly, and 'weathering', or releasing control by tuning in to the surrounding elements, BW dancers testify how agency in this practice emerges as a distributed process arising from the interconnection of the environment and the performers' dancing bodies. This practice promotes a sense of agency that occurs in relation to others, expanding beyond the individual. Through the practice of 'weathering' the body in relation to the landscape an intersubjective ecological notion of agency can emerge.

Sensing the environment, emplacing the body

The idea of 'weathering' or finding a deeper connection across the senses and the environment was significantly framed in Min Tanaka's notorious maxim: 'When I dance, I don't dance in the place, but I am the place' (Tanaka cited in Goodall 2006, 122). This idea has been discussed by Jane Goodall, who emphasizes the peculiar way of relating to spaces and places in BW, an approach that 'brings out an explicit commitment to discovering the subtle atmospheres of location' (Goodall 2006, 122).

Goodall's account resonates with Sarah Pink's approach to emplacement (2011). Pink argues that 'theoretical advances concerning the senses, human perception and place open up new analytical possibilities for understanding skilled performances and events' (2011, 344).

Pink considers the shift in the anthropology of the senses (Howes 2005) that proposes a move beyond the notion of embodiment towards the paradigm of emplacement. According to Howes, this move 'suggests the sensuous interrelationship of body-mind-environment' (Howes 2005, 7). Pink reinterprets the notion of *place* and emplacement building on the work of the philosopher Edward Casey (1993, 1996), the geographer Doreen Massey (2005) and the anthropologist Tim Ingold (2007, 2008). Such accounts challenge the idea that 'space' pre-exists a notion of 'place', and that place is more than an empirical reality people can go to and occupy. Casey, for example, has observed how the concept of *place* in Western philosophy is subordinate to the more abstract idea of *space*. The latter is often conceived of as an 'empty' space, a space that pre-exists, where instead 'place' is understood merely as the meaningful occupation of it (see Casey 1993, 1996). By calling for an embodiment and emplacement of the human subject and stressing the connections that ground mind into matter – and bodies into places – theories of emplacement aim at reinstating the primacy of 'place' over the more abstract and metaphysical concept of 'space'.

Pink stresses that the application of a theory of emplacement to what was previously understood as embodied performance offers several analytical advantages. She suggests that locating the performing body within a wider ecology can allow us to see it as an organism in relation to other organisms, thus recognizing 'both the specificity and intensity of the place event and its contingencies, but also the historicity of processes and their entanglements' (Pink 2011, 354).

Lux Eterna's work gestures towards Pink's turn from a notion of embodiment to a theory of emplacement. In presenting the director's aesthetic choices and creative approach to the production of *AURA NOX ANIMA*, I have discussed how the film is informed by a practice called *bisoku*, or moving slowly, along with 'weathering' – a practice of releasing control by tuning in to the surrounding elements. According to Lux Eterna, the practice of 'weathering' locates the performing body within a wider ecology, situating the body, grounding it into the environment. In her view, this form of 'emplacing' the senses and connecting to the weather allows the dancers to 'learn to become attuned to what the weather of your [their] body is in any given moment, in any particular environment'. In this way, 'their experiences are not simply embodied, but part of a unique environment in progress which both shapes and is shaped by their actions' (Pink 2011, 344). For the dancers in *AURA NOX ANIMA*, connecting to 'the weather of the body' is precisely what 'creates the impulse within us to move' (Lux Eterna cited in Dance Cinema 2019).

As I have argued elsewhere, BW is a performance practice that explores interconnected relationships across bodies and their environments (Pini and Deans 2021). The core of this methodology consists in understanding the body as a force of nature. In this context, agency emerges not as a feeling of being in 'control', but in an ecological sense, as a 'field of relationships', a distributed process arising from the interconnection of the performers' bodies and the landscape. Through my fieldwork I observed how the practice of BW involved becoming sensitive to the matrix of relations connecting the senses to the other performers and the surrounding environment. By directing attention towards the inner and outer 'weather' of the dancing body, BW fosters attention to the internal and external stimuli, nurturing sensitivity to the fluctuating ecology of the performance event.

Conclusion

In this chapter I have illustrated how an ecological notion of agency in performance encompasses cognitive, aesthetic and performative dimensions. By presenting some salient features that characterize BW methodology as observed and practised during a year-long fieldwork in Australia, and by addressing the creative processes behind the production of Lux Eterna's dance film *AURA NOX ANIMA* (2016), I showed how this bodily practice invites audience and performers to interrogate and explore the extent of their agency and the agency of the world towards them. As Eterna emphasized, BW consists of 'opening up and surrendering to what it is rather than what could be', prompting an exploration of new directions and perspectives stemming from kinaesthetic investigations in relation to the surrounding landscape. Through kinaesthetically questioning the agency of the performer, the environment and their mutually informing relationship, BW cultivates an ecological awareness. Agency in this context emerges as a heightened sensitivity to the internal and external stimuli coming from the body of the performer and its surroundings. As Eterna emphasized, by 'allowing my body to allow the elements to move me' (Lux Eterna cited in Dance Cinema 2019), BW performers cultivate an ecological agency.

This chapter emphasizes the multidimensional nature of embodied creative processes and how the relationship across different elements – the dancers, the director, the camera, the shared movement practice, the unstable weather, a particular environment, the elemental forces – contribute to fostering an ecological sense of agency. By revealing the entangled relationships between the various elements contributing to shaping the 'weather' of the performance event, this work stresses the relevance of tackling environment–culture–context situatedness in the study of agency and cognition in performance.[5]

Notes

1. The film *AURA NOX ANIMA* can be watched at this link: https://www.luxeterna.tv/work/4d-sf28l (Password: somapsyche).
2. I took part in several BW training and performance events organized by De Quincey Co, such as Impro-Exchange 2016. For further details, see the company's webpage: https://dequinceyco.net/research-overview/impro-exchange/
3. Tatsumi Hijikata and Kazuo Ohno are considered the founders of Butoh. Tatsumi Hijikata (9 March 1928–21 January 1986) was a Japanese choreographer, and initiator of an avant-garde dance form known as Butoh. Kazuo Ohno (27 October 1906–1 June 2010) was a Japanese dancer who, together with Hijikata, developed the dance genre called Butoh. Kazuo Ohno was internationally known for his suggestive performances. For further details on Kazuo Ohno artistic legacy and production, see the Kazuo Ohno Archives hosted by the Department of Music and Performing Arts of the *Alma Mater Studiorum* – University of Bologna, Italy: https://archivi.dar.unibo.it/files/muspe/wwcat/biblio/ohno/archives.html
4. Min Tanaka (10 March 1945–) is an avant-garde and experimental dancer and actor who trained in classical ballet and modern dance before he started his solo performance work in 1966. Deeply inspired by Tatsumi Hijikata, one of the leading figures of Butoh, Tanaka established his company and organic farm in the countryside in Japan, where he conducted dance research workshops, exploring the interconnections of bodies and environment. For further details on Min Tanaka's artistic career see Tanaka's website: http://www.min-tanaka.com/
5. I would like to thank the Body Weather community in Sydney for introducing me to this dance practice and Lux Eterna, Tess de Quincey, Linda Luke, Kirsten Packham and Gideon Payten-Griffiths for the engaging discussions. I would also like to thank Kath Bicknell, John Sutton, Kate Stevens and Amy Cook for their insightful suggestions after reading earlier versions of this chapter.

References

Blair, R. and A. Cook. (2016), *Theatre, Performance and Cognition : Languages, Bodies and Ecologies*. London: Bloomsbury Publishing.

Bresnahan, A. (2014), 'Improvisational Artistry in Live Dance Performance as Embodied and Extended Agency', *Dance Research Journal* 46 (1): 85–94.

Buhrmann, T. and E. Di Paolo. (2017), 'The Sense of Agency – a Phenomenological Consequence of Enacting Sensorimotor Schemes', *Phenomenology and the Cognitive Sciences* 16 (2): 207–36.

Candelario, R. (2019), 'Dancing the Space: Butoh and Body Weather as Training for Ecological Consciousness', In H. Thomas and S. Prickett (eds), *The Routledge Companion to Dance Studies*, 11–21, London: Routledge.

Casey, E. (1993), *Getting Back into Place: Toward a Renewed Understanding of the Place-World*. Bloomington: Indiana University Press.

Casey, E. (1996), How to Get from Space to Place in a Fairly Short Stretch of Time: Phenomenological Prolegomena', In S. Feld and K. Basso (eds), *Senses of Place*, 13–52, Santa Fe, New Mexico: School for Advanced Research Press.

Christensen, W., K. Bicknell, D. McIlwain, and J. Sutton (2015), 'The Sense of Agency and Its Role in Strategic Control for Expert Mountain Bikers', *Psychology of Consciousness: Theory, Research, and Practice* 2 (3): 340–53.

Clark, A. (2008), *Supersizing the Mind: Embodiment, Action and Cognitive Extension*. New York: Oxford University Press.

Crossley, N. (1995), 'Body Techniques, Agency and Intercorporeality: On Goffman's Relations in Public', *Sociology* 29 (1): 133–49.

Dance Cinema. (2019), 'Artist Interview: Lux Eterna', web interview. Retrieved from https://www.dancecinema.org/aura-nox-anima.html.

De Quincey, T. and I. Maxwell. (2020), 'Tess de Quincey: A Future Body', In T. Brayshaw, A. Fenemore, and N. Witts (eds), *The Twenty-First Century Performance Reader*, 166–173, London: Routledge.

Downey, G. (2010), 'Practice without Theory: A Neuroanthropological Perspective on Embodied Learning', *Journal of the Royal Anthropological Institute* 16 (supplement): S 22–40.

Downey, G., M. Dalidowicz and P. Mason. (2015), 'Apprenticeship as Method: Embodied Learning in Ethnographic Practice', *Qualitative Research* 15 (2): 183–200.

Fuller, Z. (2014), 'Seeds of an Anti-Hierarchic Ideal: Summer Training at Body Weather Farm', *Theatre, Dance and Performance Training* 5 (2): 197–203.

Fuller, Z. (2017). 'One Endless Dance: Tanaka Min's Experimental Practice', Ph.D. diss. CUNY Academic Works. https://academicworks.cuny.edu/gc_etds/2379

Fuller, Z. (2018), 'Tanaka Min', In B. Baird and R. Candelario (eds), *The Routledge Companion to Butoh Performance*, 483–90, London: Routledge.

Goodall, J. (2006), 'Haunted Places', In G. McAuley (ed.), *Unstable Ground: Performance and the Politics of Place*, 111–23, Bruxelles: P.I.E. Peter Lang.

Hallam, E. and T. Ingold (2007), 'Creativity and Cultural Improvisation: An Introduction', In T. Ingold and E. Hallam (eds), *Creativity and Cultural Improvisation*, 1–24, London: Routledge.

Haraway, D. (1988), 'Situated Knowledges: The Science Question in Feminism and the Privilege of Partial Perspective', *Feminist Studies* 14 (3): 575–99.

Howes, D. (2005), *Empire of the Senses: The Sensual Culture Reader* David Howes (Ed.). Oxford: Berg Publishers.

Hutchins, E. (2010), 'Cognitive Ecology', *Topics in Cognitive Science* 2 (4): 705–15.

Ingold, T. (2007), *Lines: A Brief History*. London: Routledge.

Ingold, T. (2008), 'Bindings against Boundaries: Entanglements of Life in an Open World', *Environment and Planning A: Economy and Space* 40 (8): 1796–810.

Jamieson, D. (2020), 'Towards an Ethics of Hybrid Agency in Performance', *Performance Research* 25 (4): 7–16.

Leach, J. and C.J. Stevens. (2020), 'Relational Creativity and Improvisation in Contemporary Dance', *Interdisciplinary Science Reviews* 45 (1): 95–116.

Lucie, S. (2020), 'Atmosphere and Intra-Action: Feeling Entangled Agencies in Theatre Spaces', *Performance Research* 25 (5): 17–23.

Marshall, J. (2006), 'Dancing the Elemental Body: Butoh and Body Weather: Interviews with Tanaka Min and Yumi Umiumare', *Performance Paradigm* 2 (March): 54–73.

Massey, D. (2005), *For Space*. London: Sage.

Merritt, M. (2015), 'Thinking-Is-Moving: Dance, Agency, and a Radically Enactive Mind', *Phenomenology and the Cognitive Sciences* 14 (1): 95–110.

Pini, S. (2020), *Stage Presence in Dance: A Cognitive Ecological Ethnographic Approach*. Ph.D. diss., Department of Cognitive Science, Macquarie University, Sydney.

Pini, S. and C. Deans. (2021), 'Expanding Empathic and Perceptive Awareness: The Experience of Attunement in Contact Improvisation and Body Weather', *Performance Research* 26 (2): 106–13.

Pini, S., D. McIlwain and J. Sutton. (2016), 'Re-Tracing the Encounter: Interkinaesthetic Forms of Knowledge in Contact Improvisation', *Antropologia e Teatro. Rivista Di Studi* (7): D.226–D.243.

Pini, S. and R. Pini. (2019), 'Resisting the "Patient" Body: A Phenomenological Account', *Journal of Embodied Research* 2.1 (2): (20:05).

Pini, S. and K. Maguire-Rosier. (2021), 'Performing Illness: A Dialogue about an Invisibly Disabled Dancing Body', *Frontiers in Psychology* 12: 566520.

Pini, S. and J. Sutton. (2021), 'Transmitting *Passione*: Emio Greco and the Ballet National de Marseille', In K. Farrugia-Kriel and J. N. Jensen (eds), *The Oxford Handbook of Contemporary Ballet*, 594–612, Oxford; New York: Oxford University Press.

Pink, S. (2009), *Doing Sensory Ethnography*. London: SAGE Publications Ltd.

Pink, S. (2011), 'From Embodiment to Emplacement: Re-Thinking Competing Bodies, Senses and Spatialities', *Sport, Education and Society* 16 (3): 343–55.

Ravn, S. (2009), *Sensing Movement, Living Spaces: An Investigation of Movement Based on the Lived Experience of 13 Professional Dancers*. Saarbrücken: VDM.

Ravn, S. (2020), 'Investigating Dance Improvisation: From Spontaneity to Agency', *Dance Research Journal* 52 (2): 75–87.

Ravn, S. (2021), 'Improvising Affectivity – Kitt Johnson's Site-Specific Performances', In S. Ravn et al. (eds), *Philosophy of Improvisation Interdisciplinary Perspectives on Theory and Practice*, 143–60, London: Routledge.

Ravn, S. and H.P. Hansen (2013), 'How to Explore Dancers' Sense Experiences? A Study of How Multi-Sited Fieldwork and Phenomenology Can Be Combined', *Qualitative Research in Sport, Exercise and Health* 5 (2): 196–213.

Sutton, J. and K. Bicknell. (2020), 'Embodied Experience in the Cognitive Ecologies of Skilled Performance', In E. Fridland and C. Pavese (eds), *The Routledge Handbook of Skill and Expertise*, 194–206, London: Routledge.

Tribble, E. and J. Sutton. (2011), 'Cognitive Ecology as a Framework for Shakespearean Studies', *Shakespearean Studies* 39(Annual): 94–103.

Wright, P. (2011), 'Agency, Intersubjectivity and Drama Education', In S. Schonmann (ed), *Key Concepts in Theatre/Drama Education*, 111–5, SensePublishers.

3

A conversation on collaborative embodied engagement in making art and architecture: Going beyond the divide between 'lower' and 'higher' cognition

Janno Martens, Ronald Rietveld and Erik Rietveld

RAAAF [Rietveld Architecture-Art-Affordances] is an interdisciplinary studio that operates at the crossroads of visual art, experimental architecture and philosophy. RAAAF makes location- and context-specific artworks, an approach that derives from the respective backgrounds of the founding partners: Prix de Rome laureate Ronald Rietveld and Socrates Professor in Philosophy Erik Rietveld.

What follows is a conversation between Erik Rietveld, Ronald Rietveld and Janno Martens, a historian of art and architecture who previously worked as an intern at the studio and as research assistant of Erik Rietveld. Starting from their own fascinations and an independent attitude, RAAAF's interventions explore possible new worlds. Through a unique working method based on multidisciplinary research with scientists and craftspeople, these interventions connect locally available social, cultural, material and natural qualities of the living environment to the past, present and future. Taking five of RAAAF's large-scale, site-specific interventions as examples, Ronald, Erik and Janno discuss how these artworks emerge from collaborative embodied engagement across multiple timescales.

Janno Martens: In a recent publication on 'ecology thinking in architecture', we related the work of RAAAF to the topic of the book through the ecological-psychological notion of affordances (Rietveld and Martens 2020). However, when talking about ecologies of collaborative skill and embodied engagement with artworks, I think it is important to understand how Erik situates these ideas within the Skilled Intentionality Framework, or SIF. Erik, could you expand on that?

Erik Rietveld: Of course. The Skilled Intentionality Framework was developed in order to connect ecological, phenomenological and neurobiological levels of analysis (Bruineberg and Rietveld 2014; Kiverstein and Rietveld 2018; Rietveld et al. 2018). It can be summarized by the following three interrelated theses:

1 There is no divide between 'higher' and 'lower' cognition. Both can be understood in terms of skilled activities of engaging with situations in the world.
2 Skilled activities are temporally extended processes in which agents coordinate to multiple relevant affordances simultaneously.
3 The affordances the environment offers are relative to the abilities available in a form of life.

We have defined affordances as relations between *aspects of the sociomaterial environment in flux* and *abilities available in a form of life* (Rietveld and Kiverstein 2014; Van Dijk and Rietveld 2017).

The form of life of a kind of animal consists of patterns of behaviour, i.e. relatively stable and regular ways of doing things (Wittgenstein 1953). In the case of humans, these regular patterns are manifest in the normative behaviours and customs of our communities. What is common to human beings is not just the biology we share but also our being embedded in sociocultural practices: our sharing steady ways of living with others. A skilled individual has developed their abilities within the dynamics of the landscape of affordances of a form of life.

The individual's intrinsic dynamics can be understood as multiple bodily states of action readiness that are attuned to the *relevant* affordances in the situation (Bruineberg et al. 2016). States of action readiness are reciprocally coupled to the landscape of affordances, in the sense that these states of action readiness self-organize and shape the selective openness to the landscape of affordances for the individual to accommodate the skilled individual's concerns, i.e. to allow them to maintain or obtain sufficient grip on the situation. In this way, some affordances in the landscape show up as more and some as less relevant to the individual's unfolding activities. Imagine attending a social event, say the opening of an exhibition: you are not just ready to see the works of art but also to interact with people. You will be attracted to the affordance of engaging with an approaching acquaintance. If you had entered there hungry, though, you would first be attracted to the affordance of grabbing some of the snacks on offer. These intrinsic dynamics of the individual thus allow for a selective openness to the *relevant* affordances (an extended hand, a snack, an artwork).

JM: And how does this relate to collaborative action?

ER: In acting skilfully one is attuned to the sociomaterial situation as a whole and for that reason there is not a clear separation between affordances offered by the material environment (a snack, an artwork) and possibilities for interaction with other people (an acquaintance). This is also a nice illustration of there being no divide between affordances for so-called 'lower' cognition (eating) and 'higher cognition' (social interaction) within SIF. In relation to skilled collaboration, i.e. coordinated social interactions, I must be *ready* for the actions of another person, and they for mine. Crucially, the nesting regularities of the shared sociomaterial environment – the familiar social context of an exhibition, a library or a diving class for example – contribute to this interpersonal attunement of states of action readiness (Rietveld and Kiverstein 2014; Van Dijk and Rietveld 2021). Some of our first work on how affordances might work at the level of expert action was concerned with how collaboration worked at RAAAF. An ethnographer observed how this skilled collaboration was ongoing even as members of the team were not physically present: architects would anticipate the preferences and responses of an absent collaborator when making decisions about details or the overall look and feel of the artwork they were creating jointly (Rietveld and Brouwers 2016). This gives an indication of how complex this kind of affordance-responsiveness is.

Because we do not differentiate between 'higher' and 'lower' cognition within SIF, no activity is excluded from being understood in terms of a skilled responsiveness to relevant affordances. The articulated goal of making an artwork can thus be understood as engagement with a large-scale affordance, and its realization approached as being sensitive and responsive to this affordance. Participant observation of how the architects at RAAAF realized an artwork over a longer period of time led us to develop a process-based account of affordances (Van Dijk and Rietveld 2021). *We understand collaborators as participants in such large-scale processes.* We found that by inviting participation, affordances can weave together to form yet larger-scale unfolding affordances:

> The process [of making an artwork] invites participants to intertwine with it and contribute their skills. They are invited to act and thus coordinate materials and transform them, so that these organized materials afford new activity to continue the process, the making of the installation. In short, the architects and other skilled individuals, familiar with architectural practices, can be invited to contribute their skills. By doing so, the larger scale process sets up the conditions for its own continuation – it forms the terms in which materials invite activity, from writing a sentence for a wall panel to seeing the opportunity to go to a store to buy carpet [to be used for constructing the artwork]. As the large-scale affordance (the new installation as a whole) thus slowly nears enactment, the range of invitations for the architects grows smaller and may become very specialized and only inviting to a very few responsive participants. By that time, anticipating the large-scale project has long made way for

the affordance of looking back on it. For others, participation has however just started, as the installation invites supported standing to the people working at the art fund, invites to be shown to visitors and [...] to be maintained and cared for in order to keep unfolding.

(Van Dijk and Rietveld 2021, 366–7)

One of the important theoretical take-aways from this study was that in the process of making, it is through an embodied engagement with the collaborative process that the work gets more determined: the complex and *large-scale affordance of creating an artwork* is constantly unfolding, to which skilled individuals respond dynamically to get more grip on the changing demands of a particular situation within this multiscaled process. Our observations suggested that finding continuity across articulated goals, written plans, images and models over time is achieved in activity. To be more precise, this continuity is achieved by coordinating activity in such a way that multiple affordances across timescales are jointly determined further. This ties into the second point I mentioned when summarizing our framework: SIF understands skilled activities as temporally extended processes, in which agents coordinate to multiple relevant affordances simultaneously.

Ronald Rietveld: Indeed, much of our work as a team consists of figuring out how to meet the demands of a particular challenge within the process of realizing an artwork. And keep in mind that our team dynamically expands as the need arises. In fact, the collaborations with various craftspeople from different fields who are not directly associated to the studio are crucial for the type of work we intend to make. Many artists like to be in control of the material aspects of a work by doing it all by themselves, which by definition limits the scope and scale of what they can make. Because we assume from the outset that we will need to collaborate with highly skilled craftspeople, we can be much freer and more ambitious in our visions for an artwork. We often do not find out what the limits and possibilities are until we collaborate with specialists who are prepared to experiment within their own craft; they too are pushed to explore what is possible. And in turn we come up with new ideas by learning about and observing how these master craftspeople engage with their materials, which is not just about seeing them work, but also includes the smells and the haptic qualities of the materials and techniques in their workshop; it is really a visceral and embodied experience of having all your senses stimulated whilst collaborating. Only through such a collaborative working method can you achieve work that truly realizes the seemingly impossible.

To give the example of *Bunker 599*: in order to cut through a Second World War-era bunker, we had to find people who had the commitment and skill to saw through several metres of reinforced concrete, which took over a month and required an enormous diamond saw. We saw what the process was and got to explore what was possible. This allowed us to push the limits even further on a follow-up project called *Deltawerk //*, where we realized a 250-metres-long land art project, for which massive amounts of concrete were cut, and then rearranged (Figure 2). We needed the experience of having worked with them earlier to get a sense of what was actually possible.

For the installation *Still Life* (Figure 3), which consists of four enormous brass plates measuring 5.3 by 3.3 metres, we were really dependent on the sole metallurgic company that was able to cast brass on such scales. It is a company that has also made the giant church bells for the Notre Dame cathedral for example, but even they had never cast brass on such a scale. The actual material properties of these giant sheets, with their rough sheen, erratic surfaces, random air pockets and other material characteristics, are really the result of an exploratory phase where we join forces with these highly skilled craftspeople who are often the only ones on a national level who can pull it off.

ER: Yes, in that sense our work can be seen as the result of a collaborative ecology that reaches far beyond the confines of the studio, which is actually rather small. As we observed in our ethnographic study of realizing an artwork within the studio (Van Dijk and Rietveld 2021), it is through our expertise with involving specialists from the beginning that we are able to anticipate the direction that the process is taking. The more such participants are invited to contribute their skills to the process, the more direction it can take, and the more its participants will be able to attune to the direction of its large-scale unfolding (cf. Noë 2012, 25 ff.).

FIGURE 2 *RAAAF | Atelier de Lyon – Deltawerk // (2018). Photo by Jan Kempenaers. © RAAAF.*

Deltawerk // *brings into question the Dutch effort to realize indestructible sea defences in times of climate change and rising sea levels. At the same time, it is an experiment in the active creation of ruins and a plea for a radically different approach to cultural heritage. By digging out a colossal wave basin that served as test site for the Dutch Delta Works between 1977 and 2015, this monument of the Dutch battle against the sea is suddenly inundated. By sawing through the concrete flume and re-arranging its parts, a new rhythm of slumping slabs reveals the true size of this massive laboratory. It allows visitors to walk over the water and into the flume, confronting them with a perspective on the void that now inhabits the space between the heavy slabs.*

FIGURE 3 RAAAF – *Still Life (2019)*. Photo by Jan Kempenaers. © RAAAF.

During the Cold War, millions of bullets were made for NATO soldiers worldwide in the former bullet factory The Hem. At the time, the factory was full of trays with brass bullet casings. The artwork Still Life *brings the abandoned war factory's history into question and creates a link between the present, past and future of this historically burdened heritage. The source material of the bullet production has been melted and cast into four heavy brass plates. The large plates move in between the columns in an unpredictable rhythm; together they open and close one's perspective on the immense space. Their movement forces the visitor to relate to the work over and over again. The brass plates move slowly away but inevitably return.*

Not all of these collaborations are with 'hands-on' crafts such as casting metal or cutting concrete. Sometimes the opposite is the case, for example with our proposal for a giant block of sand (Figure 4). For this project we are collaborating with materials scientists at the Technical University of Delft who are researching ways to use microbes in order to turn sand into sandstone. This is a highly experimental technology, and they have yet to pull their method out of the infamous 'valley of death': the phase of product development where many ideas fail to be implemented because it

FIGURE 4 *RAAAF | Atelier de Lyon – Sandblock (2019).* © RAAAF.

is too difficult or costly to scale up to commercial applications. By using their experimental technique on our experimental artwork, we are engaged in a mutually beneficial collaboration between their expertise and our vision. Actually, on an ecological level, it might even be considered as a collaboration between them, us and the microbial organisms that turn the sand into sandstone. This goes to show how collaborative engagement is situated in a rich landscape of affordances and is highly dependent on the variety of abilities that are inherent to different forms of life, as I mentioned in summarizing SIF. In this case, the abilities we depend on for a work come from the forms of life of us as artists, the materials specialists as experimental scientists and the abilities of microbial organisms when they are presented with the right environmental conditions.

JM: Besides skilled craftspeople and material scientists, there is a third category of specialists you often collaborate with: cultural historians. They obviously bring something to the table as far as research goes (they helped identify different types of bunkers for the *Bunker 599* project for example), but perhaps their role in the actual materialization of the artwork is far more limited.

Speaking about (cultural) history: what I have found to be interesting since first encountering your work is the way it relates to historical precedents. Within architecture, many others have held similar ideas with regard to creating affordances – whether they be for creating places to sit or are geared towards larger social phenomena. For example, to me, the historical similarities between a RAAAF project such as *Trusted Strangers / New Amsterdam Park (N.A.P.)* (Rietveld et al. 2019) and the ideas of Jane Jacobs (1961) or Jaap Bakema (cf. Van den Heuvel 2020, 19) initially seemed very obvious. Especially the idea of being able to observe others is a very recognizable tenet that has long been championed as a way of realizing urban social cohesion. However, after becoming more acquainted with SIF, I did notice some differences in how these principles were thought to work back then and how they are conceptualized by Erik and his colleagues. Whereas Jacobs or Bakema regarded the idea of 'eyes on the street' as enabling some sort of basic connection with public life, in SIF it becomes more about *learning* to be exposed to one another, of a *process* of becoming part of community that involves a certain skill: you refer to the notion of 'bi-cultural competence' (cf. Voestermans and Verheggen 2013) as something that can be learned, i.e. as a skill. I feel this is a more dynamic approach to social cohesion than the ideas from the 1960s, which generally related to the community as a whole but to a certain extent neglected individual performance. Erik, could you say a bit more about how individual performance relates to collaborative performance?

ER: As for the references behind *Trusted Strangers* (Figure 5): this project was mainly informed by a study of social cohesion in a multicultural district of Amsterdam (Nio et al. 2008). The notion of 'trusted strangers', of being exposed to one another, was partly based on this. But other elements were grounded in ecological-enactive insights. One of the key aspects of the park is that all spaces are to be freely accessible to the public, which ensures that visitors can roam freely. This allows people to over time explore more and more aspects of the park.

With regard to individual versus collective performance, it is important to note that each individual grows up in a multitude of practices and contributes to the maintenance of a collectively shaped landscape of affordances through having their development sculpted by other members of these practices. However, each growing and learning individual takes their own unique and particular path to do so – a path that is shaped by and shapes the skills, sensitivities and current concerns of that individual. Thus, in any concrete situation the field of relevant affordances forms at the point (or rather the line) where both the individual's path and the landscape intermingle and together develop further. As they make their way together, neither remains unchanged (cf. Ingold 2011, 2018). For example, the invention of diving equipment enriches the landscape of affordances, which can in turn enable the formation of new sociocultural practices such as deep-sea diving, and further improvements of the tools that support it. There is an aspect of dynamic change which concerns both the individual and the collective as well as the landscape of affordances.

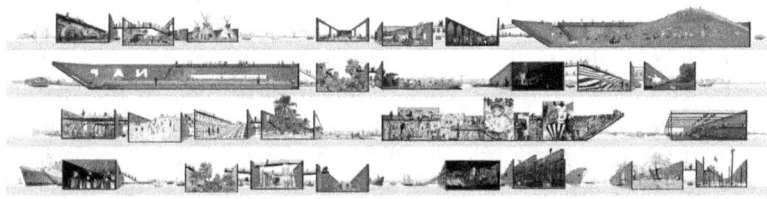

FIGURE 5 *RAAAF|Atelier de Lyon – Trusted Strangers | New Amsterdam Park (N.A.P.).* © *RAAAF.*

Taking into account the urgent need for a good public domain along with our own research into and views on how to create such spaces, RAAAF and Atelier de Lyon have proposed a temporary floating park called Trusted Strangers | New Amsterdam Park (N.A.P.). *Along the northern bank of the River IJ, a grid of twenty-four large barges will shelter a hidden water world: the basis of a new floating park. By accommodating both subcultural niches as well as 'public' activities with a broad appeal, the park becomes a condensed city floating on the water with an abundance of social affordances. And because these affordances are made to be attractive for people with diverse sociocultural backgrounds, they are able to generate new patterns of behaviour and invite surprising spontaneous interactions. The notion of 'trusted strangers' – the importance of people becoming 'familiar strangers' to their neighbours in order to bolster social cohesion (Blokland-Potters 2005) – served as the project's premise: observing and being observed is made possible by the material environment (portholes, cut-throughs and meandering overhead pathways all contribute to this) and is essential to the culture of this park.*

JM: While *Trusted Strangers* is very collective in nature and affords distinct and particular activities – it is a park after all – many other interventions by the studio are much more ambivalent as to what kind of (re)actions they could or should evoke. For much recent work by RAAAF, which relies strongly on poetic and abstract qualities, the collaborative aspect is not quite as clearly present. In those cases, the performance of the artwork seems more related to the personal embodied experience rather than to collaborative engagement. How do you conceive of this relation between personal experience and collaborative performance of a work – embodied or otherwise?

RR: In each project, we make sure both that the visual image is striking and that the work allows for a certain immersive embodied experience of the work; a 'total experience'. From the very outset we consider what kinds of engagement are afforded by the spatial and material aspects of a work, and without the engagement of the visitor a work would definitely not be 'complete'. Usually it is at the level of the 'total experience' that people initially relate and react to the work: to the scale, the materiality, the experience of being immersed in something. You could see this as a collaboration between the work and the visitor. However, as you know we always make context-specific work and that means that we also often convey a certain cultural history, or want to bring something to the attention, or to change particular practices such as conservation policy (Rietveld and Rietveld 2017). It is in the latter sense that we characterize our interventions as 'strategic'. We have recently begun to conceptualize these different aspects as the 'inner horizon' and the 'outer horizon' of a work (cf. Merleau-Ponty [1945] 2002; Rietveld and Rietveld 2020). When we had to communicate our project to others, this would usually be done through photos of a work in its context, and we would then follow up by telling its 'backstory'; with relating the work to the cultural historical and societal references that informed it. But it was very difficult for people to relate this 'outer horizon' to the 'inner horizon' of the work, which is often constituted by embodied experience related to the materiality of the work. So now we've begun to make very detailed composites of hundreds of high-resolution close-up photos of our work in order to convey their materiality and experiential qualities, to translate them to an exhibition setting.

ER: The division between 'inner' and 'outer' horizon is an analytic distinction, because in the end the idea of a total experience is that it allows for both of these horizons to be present at the same time when one actually visits the work.

JM: With some of the work that is strongly related to cultural history, which *Bunker 599* for example certainly is, I guess this experience would indeed amount to a type of collaborative engagement because it conveys history in a very tangible and embodied manner. It is very rare to be engaged with history in a non-linguistic manner, but with you being specialized in embodied experience and action, it makes sense that this would be the way that the work of RAAAF allows for that. I am inclined to say that collaborative performance is in fact also a part of the more autonomous works, though not so much in the relation between different simultaneous visitors but rather in the relation between the (socio)material and historical environment and people's experience, of what it lays bare, and of how it allows for a visitor to experience a particular place in a new way. Having said that, I do wonder what the role of language is within SIF, and how it relates to (creative) collaboration.

ER: As far as language is concerned, it is important to keep in mind that although my academic work (and by extension the work of RAAAF) has tended to focus on embodied experience and skilled action, language itself is certainly not something that is outside of its scope. Quite the opposite, crucially: because we do not distinguish between 'higher' and 'lower' cognition, language can also be understood as skilled engagement with affordances and a type of collaborative embodied action (Kiverstein and Rietveld 2020). The materials from which speech is made – expressive bodily activities – take form through the regular ways of acting of the members of the linguistic community. The expressive possibilities available to speakers of a language – what Merleau-Ponty (1945/2002) called 'spoken speech' – are sustained by the regular, habitual patterns of talking. These established ways of speaking lay out what makes sense, and what does not, in the language-speaking community to which the individual belongs. If a person is to speak and make themselves understood, it will only be by acting in ways that fit with the patterns for doing things already mapped out in the standing practices (cf. Wittgenstein 1953). The regular pattern of doing things is essential because it is relative to this agreement in how to take part in the practice that evaluations can then be made as to whether a use of a word in an utterance is appropriate, or inappropriate, correct or incorrect (Rietveld 2008; Van den Herik and Rietveld, 2021).

JM: This is fascinating to me, because I recently studied a case of two major names in art history which showed that skills in relating to a foreign linguistic community can make or break a career (Martens 2020). One was able to be sensitive to the established ways of speaking, as you call it, whilst the other was not. Now, I might be on a bit of a tangent here, but this notion makes me think of a prerequisite for collaboration at RAAAF that we have not touched upon yet: the notion of someone 'fitting in' with the team, which in my experience of working there has always been very important. It seems to resonate with what Erik just said about established ways of speaking: in order to understand the work being made, and understand each other for that matter, one needs to have a feeling and sensitivity for what makes sense within the 'community' of the studio. While this does not immediately relate to any type of specific skill, it nevertheless seemed important for successful collaboration. Ronald, what are your thoughts on this?

RR: Not only is it important, having somebody fit in with the team is a matter of pure necessity. We are a small team, all working at the same table, so naturally it is important that everybody gets along. It is true that this is not a matter of skill or affiliation with just the work alone: those are prerequisites that go without saying. But whether somebody fits in with the team or not concerns qualities that have nothing to do with that; in fact, I would argue that in this regard it is more important what kind of Spotify playlist somebody listens to than what work they have in their portfolio. People have to be able to relate to the motivation of why we make the type of work we do and share some of the deeper fascinations that drive these – if you do not love concrete you don't stand a chance! At this level, completely different factors are important for successful collaboration, such as a shared sense of humour or music taste.

JM: I think those remarks constitute a very suiting conclusion to a conversation related to collaborative performance: music and humour are certainly important within any cognitive ecology as far as I am concerned! Thank you both for this insightful conversation.[1]

Note

1 We thank Joel Krueger, Ian Maxwell, John Sutton and Kath Bicknell for their helpful comments. Financial support was provided by an ERC Starting Grant for the project AFFORDS-HIGHER (679190) and a NWO VIDI grant awarded to Erik Rietveld by the Gieskes-Strijbis Fund and the Mondriaan Fund with a Stipendium for Established Artists awarded to Ronald Rietveld, and by the FWO with a PhD fellowship grant (1143521N) awarded to Janno Martens.

References

Blokland-Potters, T. V. (2005), *Goeie buren houden zich op d'r eigen: Sociale relaties in de grote stad*. Den Haag: Gradus-Hendriks Stichting.

Bruineberg, J., J. D. Kiverstein and E. Rietveld (2016), 'The Anticipating Brain Is Not a Scientist: The Free-Energy Principle from an Ecological-Enactive Perspective', *Synthese 195* (2018): 2417–44.

Bruineberg, J. and E. Rietveld, E (2014), 'Self-Organization, Free Energy Minimization, and Optimal Grip on a Field of Affordances', *Frontiers in Human Neuroscience 8* (599).

Ingold, T. (2011), *Being Alive: Essays on Movement, Knowledge and Description*. Abingdon: Routledge.

Ingold, T. (2018), *Anthropology and/as Education*. Abingdon: Routledge.

Jacobs, J. (1961), *The Death and Life of Great American Cities*. New York: Random House.

Kiverstein, J. and E. Rietveld (2018), 'Reconceiving Representation-Hungry Cognition: An Ecological Enactive Proposal', *Adaptive Behavior 26* (4): 147–63.

Kiverstein, J. and E. Rietveld (2020), 'Scaling-up Skilled Intentionality to Linguistic Thought', *Synthese*. doi. org/10.1007/s11229-020-02540-3

Martens, J. (2020), 'Lost and Found in Translation: The Postwar Adaptation Strategies of Sigfried Giedion and Alexander Dorner', *Journal of Art Historiography 23*: 1–21.

Merleau-Ponty, M. (2002), *Phenomenology of Perception*, trans. C. Smith, London: Routledge. (Original work published 1945)

Nio, I., A. Reijndorp and W. Veldhuis (2008), *Atlas Westelijke Tuinsteden Amsterdam: De geplande en de geleefde stad*. Haarlem: Trancity.

Noë, A. (2012), *Varieties of Presence*. Cambridge, MA: Harvard University Press.

Rietveld, E. (2008), Situated Normativity: The Normative Aspect of Embodied Cognition in Unreflective Action', *Mind 117* (468): 973–1001.

Rietveld, E. and A. A. Brouwers (2016), 'Optimal Grip on Affordances in Architectural Design Practices: An Ethnography', *Phenomenology and the Cognitive Sciences 16* (2017): 545–64.

Rietveld, E., D. Denys and M. van Westen (2018), 'Ecological-Enactive Cognition as Engaging with a Field of Relevant Affordances: The Skilled Intentionality Framework (SIF)', In A. Newen, L. De Bruin, and S. Gallagher (eds), *The Oxford Handbook of 4E Cognition.* 41–70, Oxford: Oxford University Press.

Rietveld, E. and J. Kiverstein (2014), 'A Rich Landscape of Affordances', *Ecological Psychology 26* (4): 325–52. doi.org/10.1080/10407413.20 14.958035

Rietveld, E. and J. Martens (2020), 'Architecture and Ecological Psychology: RAAAF's Explorations of Affordances', In D. van den Heuvel, J. Martens, and V. Muñoz Sanz (eds), *Habitat: Ecological Thinking in Architecture*, 128–35, Rotterdam: nai010.

Rietveld, E. and R. Rietveld (2017), 'Hardcore Heritage: Imagination for Preservation', *Frontiers in Psychology*, 8: 1995.

Rietveld, E. and R. Rietveld (2020), *The Landscape of Affordances.* Amsterdam: Black Paper Press.

Rietveld, E., R. Rietveld and J. Martens (2019), 'Trusted Strangers: Social Affordances for Social Cohesion', *Phenomenology and the Cognitive Sciences 18* (1): 299–316.

van den Herik, J.C. and E. Rietveld (2021), 'Reflective Situated Normativity', *Philosophical Studies 178*: 3371–3389.

van den Heuvel, D. (2020), 'Habitat and Architecture. Disruption and Expansion', In D. van den Heuvel, J. Martens and V. Muñoz Sanz (eds), *Habitat: Ecological Thinking in Architecture*, 8–21, Rotterdam: nai010.

van Dijk, L. and E. Rietveld (2021), 'Situated Anticipation', *Synthese 198* (1): 349–71.

van Dijk, L. and E. Rietveld (2017), 'Foregrounding Sociomaterial Practice in Our Understanding of Affordances: The Skilled Intentionality Framework', *Frontiers in Psychology 7*: 1969.

Voestermans, P., and T. Verheggen (2013), *Culture as Embodiment: The Social Tuning of Behavior.* Hoboken: Wiley-Blackwell.

Wittgenstein, L. (1953), *Philosophical Investigations.* Oxford: Blackwell Publishing.

Commentary: Redirecting our telescope

Amy Cook

In 1611, Galileo asked spectators to look at the heavens through his telescope and see the moons of Jupiter and the mountains on the moon. They stood on a rooftop as he pointed the instrument towards the sky and when it was their turn to look, they were unimpressed. Where he saw planetary objects in space, they saw variations of grey and white. They turned the telescope towards the earth and were dazzled to discover they could make out a far-away building. They could see that. The mountains on the moon made no sense (Spiller 2000). The essays in this section remind me of this story: turning our gaze in a different direction to where the actual discovery is. More than the new view, however, is the story that goes along with it: where before we looked towards the dancer, the actor, the building, now, they tell us, look to the whole.

Evelyn Tribble turns our gaze to the critical, interweaving ecosystem that allows us to keep watching the stage. Tribble taught us to think differently about the 'genius' of Shakespeare in her book *Cognition in the Globe: Attention and Memory in Shakespeare's Theatre* (2011), wherein she argued that what we attributed to one man was the work of a complicated ecosystem that created and scaffolded genius. Without the plots backstage, the structure of the playhouse, the apprentice system, the poetry, for example, what came to be known as *Hamlet* could not have been. We wanted a sole, male genius, so we found one. This is not to indulge in the authorship debate, but rather to initiate a conversation about the nature of authorship in theatre. Like the frontispieces of the quartos that do not recognize Shakespeare as the author but rather locates the authority to the players: 'as it was recently performed by Lord Chamberlain's Men,' Tribble locates the genius in the ecosystem that generated the plays.

Here Tribble turns our attention from the virtuosity of the actors to the exquisite dance required to keep our attention onstage at Shakespeare's Globe. She broadens her view to the wider cognitive ecology which she defines as 'the physical, material, social, affective and mental resources that sustain and maintain' the performance in question. For Tribble, this includes the many people necessary to keep the patrons safe and behaving. After each performance there is a report written about what went on in the rest of the theatre while the actors took their entrances and their exits. It turns out, quite a lot happens, and Tribble mines this archive of fainting and vomit to depict a practised but spontaneous teamwork. When 'help' is offered from the star – the seeming conductor or controller of the space – chaos ensues because the collaboration of the team was interrupted by a hierarchy it evolved to do without. From centre stage, he seems to lead the troops to calmly take care of a patron in distress, but his illusion of control led to a challenge to the group's actual – though less visible – control.

If I'm honest, I feel this way most days. Over morning coffee, I think about my day, the work I will get done and the austerity I will impose on my system: light lunch, no afternoon sugar, less evening alcohol. But my microbiome, my front-of-house team of organisms that make up a majority of my cells, have already started a chemical process that will lead me through my day – wanting some of the very things I had instructed the system to do without. This is not to say that I have no control, but my fantasized individuality and autonomy, should be reconsidered in light of my minority stake in what I consider 'me.' If most of 'me' is something else, then I deserve a lot less credit, and blame, for the things I do. It's time for the same kind of identity reassessment in where we look to find skill. It's not where we were looking before.

Like Tribble's chapter, the others beautifully redirect our telescope. The artists/architects of RAAAF (Rietveld Architecture-Art-Affordances) are intentional in their prioritization of collaboration. They intentionally see themselves in collaboration with the tradesmen that pour the brass, the stone cutters that cut a path through a Second World War bunker and the organisms that produce the work. Collaborators are united by a process and ethic – and, as Ronald Rietveld says, 'if you do not love concrete you don't stand a chance.' RAAAF creates spaces that perform, that afford different social experiences. They think a lot about the experiences that are baked into our environments and how they might change them. Janno Martens, Ronald Rietveld and Erik Rietveld discuss how RAAAF's interventions are the result of collaborative embodied engagement across multiple timescales.

They might see my craving for an afternoon cookie as part of the kind of 'lower cognition' that is generally disregarded but need not always be 'below' the 'higher' cognition that calculates the calories in the treat. Indeed, their work with materials like concrete, brass and sandstone depends upon and displays the processes necessary to produce the material. Erik Rietveld sees his work as involving all of it: 'Actually, on an ecological level, it might even be considered as a collaboration between them, us and the microbial organisms that turn the sand into sandstone.' We might build different spaces, prioritize different collaborators or notice new things if we shifted our perspective on what counted as 'thinking' or 'working.'

Structured as an interview, this chapter invites us to imagine the theoretical work as collaborative as well. The voices come together like flaneurs coalescing around one of their public art works, engaging with the space anew and through doing so, theorizing its importance. I find this invitation incredibly important, as the separation between artist and scholar is unhelpful. I co-taught a course once with an actor/director professor who constantly referred to herself as a 'worker bee' and thus somehow less interested in, or capable of, the kind of 'thinking' that I would bring into the classroom. How irritating this was to me: she created theatre that generated thinking in the cast and the audiences and then pretended she wasn't the intellectual. Why do we think it matters where or what the thinking is? I get angry when people accuse me of 'overthinking' something, as if that's a thing. I want to take the diamond saw that RAAAF used to cut apart the Second World War bunker and destroy our concept of 'thinking' so we are forced to find new language to describe the generative, collaborative, embedded, distributed work that happens when a thing of beauty or great use or virtuosic skill is created.

Sarah Pini adopts a cognitive, ecological and phenomenological perspective in exploring the work of a dance company from both inside and out. Like RAAAF rethinking thinking, Pini's work suggests that we should interrogate agency. Where does movement come from? Does the dancer dance or does the environment dance the dancer? Pini explores a dance company that dances with and in the environment, seeing the art where the difference between dancer and space blur. The process and the product focus on the cultivation of the relationship with the environment, as if dancer and space can become one, as that which impacts the movement of the sandy dunes also moves the inside and outside of the body. This dance methodology, Body Weather, incorporates the elements – from the body to the camera to the 'elemental forces' – through stillness and slowness. Again, the work insists that we alter our perspective – what's figure and what's ground – to see what moves and what's still, what dances and who looks.

If, as they say, the dance of Body Weather cultivates an 'ecological awareness', we might expand our sense of self to include the forces that move us – whether from within or without. These dancers are trained to slow way down, and this causes me to wonder if the typical 'idle speed' of humans allows us to pretend that we are the locus of agency. Do we move too fast to notice we are not the initiators of our movement? Pini refers to their training as 'weathering' – 'Through the practice of "weathering" the body in relation to the landscape an intersubjective ecological notion of agency can emerge.' This is how we should begin to think about collaboration, skill and 'thinking'. It happens at a timescale that is beyond what we are capable of imagining. What would it look like if I thought that any idea I had must first be weathered – like the rock in the stream that does not complete its mission until it is smooth and then small and, finally, sand. I admit that for as long as it takes me to write a paper or a book or even this reflective essay, I at least believe that I will live to see the idea out. If, instead, I knew that anything truly valuable would require the weathering of years, generations, epochs, I might surrender to the sand dunes, like Pini's dancers, secure in the knowledge that it's not for me to assess the impact.

I love the experience of shifting the perspective from figure to ground, from data to noise. I had thought the goal was to understand that which I was studying, and I realize I was looking at the wrong thing or at the wrong timescale. The essays in this section help me turn my attention, to look anew at what I thought I understood.

References

Spiller, E. (2000), 'Reading through Galileo's Telescope: Margaret Cavendish and the Experience of Reading', *Renaissance Quarterly 53*: 192–221.

Tribble, E. B. (2011), *Cognition in the Globe: Attention and Memory in Shakespeare's Theatre*. New York: Palgrave Macmillan.

PART TWO

Learning, collaboration and socially scaffolded cognition

4

'No elephants today!' Recurrent experiences of failure while learning a movement practice

Kath Bicknell and Kristina Brümmer

Meredith [teacher] announces that we're about to finish the warm-up.¹ I'm tired already and checking the time. 45 minutes to go. Meredith tells us to pair up with a partner to move through the circuit of exercises. Kath and I start at a station where we are meant to lift into an L-shaped handstand while using a pile of thick soft mats to support our feet. This is tough for me today. Meredith and Kath both tell me that I'm not handstanding straight and I'm hollowing my back. My arms and shoulders start burning and I feel frustration rising in me. Last week, this exercise felt much easier, and in general, I enjoyed class much more. Today I just wish to go home; my body feels heavy and immobile. [...] We proceed to the next station where we're supposed to cartwheel from a vault box to dissolve our handstands. Kath encourages me to give it a try first. The move feels odd and – accompanied by a loud thud – both of my feet crash on the floor. I look at

Kath who's smiling but looking sceptical. I feel embarrassed about my collapse and I'm happy that I have her here to share that experience. She helps the learning process by making me calm with her composure. Jokingly, we say that maybe we are elephants.

KRISTINA BRÜMMER, FIELDNOTE, HANDSTAND CLASS, 28 MARCH 2019.

Kristina lands the cartwheel out of handstand move.
Kath: 'No elephants today!'
Kristina: 'No elephants!'
I try to cartwheel out on my less dominant side.
Kath: 'A small elephant! I have an elephant on this side, but not the other side.'
Kristina: 'Mine's on this side!'

KATH BICKNELL, FIELDNOTE, HANDSTAND CLASS, 11 APRIL 2019.

Introduction

These fieldnotes come from an exploratory ethnographic study. In 2019, the authors of this chapter, Kristina and Kath, undertook weekly 'Handstands Foundation' classes at a movement school in Sydney, Australia, for six weeks. Our motivation to join the class was academic and personal alike. Experienced in other movement and gymnastic practices, both of us were eager to try a new workout together in our free time. But as it was a two-month research stay that brought Kristina to Sydney, we also intended to render our handstanding experiences productive for theoretical reflections on the learning of movement practices. We picked the class due to its agreeable timeslot and location and without much knowledge about its organization. The class was open to learners of every level of proficiency and bookings could be made on a casual basis. The classes aimed to teach, as the school's website put it, 'foundations to achieve a strict handstand'. The basic version of the strict form we learnt required us to align all body parts in a straight posture: wrists, shoulders, pelvis and feet must form a continuous line for the body to be able to stand upside down for a certain amount of time. Exercises included training drills to strengthen and condition our wrists, shoulders, core, overall bodily awareness and our sense of balance and alignment.

In this chapter, we do not tell the story of how we successfully learned a 'strict handstand'.[2] After six weeks of classes and occasional practice between them, we were both able to hold an unsupported handstand for a few seconds, but only sometimes. We found cyclical experiences of success and regression to abound in learning. This is unsurprising to most people who have tried to learn a new movement practice. By contrast, academic studies of skill learning and concepts of practices tend to emphasize ability, improvement and mastery (Bicknell 2021; Boll and Lambrix 2019). Elaborating on the 'elephant episode' sketched in the fieldnotes above empirically and theoretically, we focus on how two novices, ourselves, experienced and dealt with fluctuation and failure when learning a movement practice. Limiting our analysis to data from this initial fieldwork, our aim is to highlight the beneficial roles of objects, learning exercises and other people in the learning context – the ecology of skill – in encouraging novices to work with failure, rather than fearing it.

Integrating sociology of practice and cognitive theory

Our theoretical approach is informed by two different yet compatible perspectives: approaches from the sociology of practice and cognitive science that share an interest in studying human activities (including learning) as embodied performances. Drawing on sociological practice theories, we conceive of a strict handstand as a particular cultural practice. By 'practices' we mean, following Schatzki (2002, 72), historically evolving, recognizable forms of activities or 'bodily doings' characterized by certain demands and requirements as well as specific socio-material arrangements which provide the context for their performance. The exact forms of practices can be more or less rigidly defined. While many everyday practices allow for a certain variability of performance, others (such as sports practices) are strictly codified, defining precisely which bodily doings count as functionally and aesthetically adequate realizations. From a practice sociological perspective, learning a handstand in a movement school is a process of learners incorporating the idealized form of the practice in a specific socio-material context arranged for this exact purpose.

Due to an increased interest in mechanisms of social order and reproduction, there is a dominant tendency in sociological practice theory to view practices as routine activities. According to Bourdieu's approach, practices proceed smoothly via practitioners' embodied practical knowledge or 'practical sense' (Bourdieu 1990, 103) and are carried by 'able' bodies that learn to adhere to a practice's requirements and demands through habitualization (Bourdieu 1990, see also Reckwitz 2002). In his late work, the *Pascalian Meditations* (2000), however, Bourdieu relativizes the equation of practices and routines by talking about moments of failure. Using the example of a tennis player performing a stroke gone wrong, he writes:

> [...] habitus has its 'blips', critical moments when it misfires or is out of phase: the relationship of immediate adaptation is suspended, in an instant of hesitation into which there may slip a form of reflection which has nothing in common with that of the scholastic thinker and which through the sketched movements of the body [...] remains turned toward practice and not towards the agent who performs it.
>
> (Bourdieu 2000, 162)

Bourdieu describes 'practical reflexivity' as a form of movement control operating in moments of failure. He introduces it not as a cognitive rationality or conscious thought, but as an ability embedded in practice. This allows practitioners to monitor, critically evaluate and correct their own doings drawing on embodied knowledge of the practice performed as well as a kinaesthetic 'feel' for and awareness of what went wrong.

The concept of practical reflexivity seems instructive for gaining a better understanding of the elephant episode portrayed at the beginning of this chapter. But as Bourdieu only mentions practical reflexivity once, the concept remains vague. We now present ideas from cognitive theory that enable us to better grasp the forms of reflexivity and movement control that Bourdieu only hints at.

'Hybrid' theories of skill detail the intertwined roles of cognition and motor action during skilled movement and are a rapidly growing area of interest in cognitive science (see Pacherie and Mylopolous 2021 for a recent overview). Alongside experientially informed accounts of skilled movement (Sutton and Bicknell 2020), hybrid skill theorists provide evidence against influential models of skilled action that have long emphasized a lack of cognition in expert movement practices (e.g. Anderson 1982; Fitts and Posner 1967). Christensen et al. (2016), for example, describe the *productive* roles of cognition in experts' experiences of movement practices, which include adapting to novel or challenging circumstances. These authors claim their 'Mesh' theory of skill 'develops further a theoretical explanation for the persistence of cognitive control in advanced skill, and characterises a transformation in cognitive control to more efficient forms that involve substantial non-linguistic structure' (p. 63). Cognitive control, in this view, incorporates embodied, situated awareness via a range of multimodal sources and the ability to use such awareness to guide movement. One example of strategies for movement control used by advanced practitioners are techniques Sutton (2007) describes as instructional nudges, cues or compressed practical talk. These may be substantially reduced forms of longer utterances that guide movement, or distinctly nonverbal, such as sights, sounds, sensations and touches (Bicknell 2011; Brümmer and Alkemeyer 2017; McIlwain and Sutton 2014).

Initially motivated by debunking the notion that cognition is largely absent in expert experiences of doing, hybrid skill theorists have focused on advanced experiences of skill. We suggest that to better understand how movement control *transforms* into efficient forms in experts, further attention is needed on how these processes develop in the first place. After all, the early stages of learning are characterized by novel and challenging circumstances. The development of abilities for dealing with such circumstances is often crystallized in moments of failure, regression, near-misses and surprise success.

Combining concepts from sociology of practice and cognitive science provides a fuller picture of learning a movement practice than either approach alone. The detail and emphasis on the roles of cognition in adapting skilled movement to suit unique challenges and contexts provide a theoretical foundation for exploring strategies that help not only experts but also novices deal with, or respond to, the unstable moments, or blips, in practices which Bourdieu alludes to but does not develop. Integrating Bourdieu and Mesh, we understand movement control as an embodied, multimodal ability of practitioners to monitor, (critically) evaluate and guide their own bodily doings. Careful consideration of the socio-material contexts in which the learning of a movement practice is situated, meanwhile, can promote a clearer understanding of how movement control is developed and scaffolded.

A collaborative ethnographic study of f(l)ailing at handstands

Our analysis of how novices experience and deal with failure feeds from empirical data we gathered in a 'short-term' ethnographic study (Pink and Morgan 2013) in which we investigated the learning and performing of a movement practice in its 'natural habitat': a movement school. By focusing on learning as it actually happened, our study provides a much-needed counterpoint to abstract theoretical studies or modified, experimental task environments. These are dominant approaches used to study skill learning in cognitive, exercise and sport science as well as the philosophy and psychology of skill. Compared with other ethnographic studies on sport and movement learning, and ethnographic approaches emphasizing longer-term immersion in the learning context (e.g. Downey et al. 2015; Wacquant 2004), the six weeks of our participation in class is a short time span. Given our interest in the very early stages of learning, the short-term approach was appropriate. We augmented this approach through two ethnographers working collaboratively to gather data, a process that enabled us to combine and compare observations and perspectives. Our exploratory study does not cover the entire process of learners becoming able to perform a 'correct' handstand but instead examines events within this timeframe to highlight the roles of fluctuation and failure in the very early stages of learning.

In the first three weeks of class, we engaged in open, unfocused participatory observations. We each sat down immediately after class to note our impressions and perceptions in independent fieldnotes. We chatted occasionally about what had happened in class, but we only began to jointly analyse our writing in week four. We exchanged our fieldnotes and identified some common themes. We both wrote repeatedly about moments of success and failure: of progress and regress, and the cyclical feeling of being 'in control' of our bodies sometimes, while our bodies felt obstinate at others. We therefore focussed on the roles of failure in supporting success in the remaining classes and developed the 'elephant episode' for this chapter as it provided an excellent example for further exploring the relationship between failure and success.

Empirical analysis

In the following sections, we analyse the events that informed the fieldnotes at the beginning of this chapter. Inspired by practice sociologies, we first look at the socio-material arrangements of the exercises in which classes were organized and our learning was situated. Next, we describe the spontaneous development of 'being/ having an elephant' as a diagnosis of failure in the first situation and as a cue for landing lightly in the second, before examining the rich facets of this short expression as a multimodal form of movement control. Lastly, we discuss why it is crucial and fruitful for concepts of (learning) movement practices to take into account moments of failure, regress and bodily obstinacy and suggest conceiving of learning as a non-linear process.

Socio-material arrangements of the handstands classes

The handstand classes involved two teachers, Meredith and Josh, as well as approximately twelve learners. Classes were structured according to a recurring pattern: for the first fifteen minutes, everybody gathered for a warm-up to prepare for the following forty-five minutes of exercise. The exercises designed for incorporating the 'strict' form we were learning took place at different stations involving diverse artefacts as supporting structures: floor mats, thick soft gym mats, handstanding chairs, vault boxes, walls. They provided a means for helping us develop the several physiological components and systemic adaptations that are needed to support an effective handstand (Downey 2012), bit by bit. The teachers usually gave a short introduction to each exercise. While Josh performed them, Meredith provided a verbal explanation and used Josh's body as an object of demonstration, pointing to certain parts that the learners should pay special attention to while executing the respective exercises: push through the tips of the fingers and the palms to avoid collapsing into the wrists, knit together between the bottom of the rib cage and the top of the pelvis to avoid arching the lower back. This procedure echoes Goodwin's (1994) observations about the role of experts educating the attention of those as-yet-unfamiliar with the subtleties of a practice. Using the term 'professional vision' to describe the basis as well as the goal of such a process, Goodwin highlights the phenomenological salience the procedure promotes regarding what a learner may observe in the learning environment. Through increased awareness of fingertips and palms, and of abdominals and backs, among other muscles, regions and joints, each exercise in the handstand class educated our attention to kinaesthetic sensations and shapes according to the demands of the aimed-at practice.

Within the handstands classes, arrangements of artefacts further educated our attention and isolated specific body parts and muscle groups. They reduced the complexity of handstanding and guided our awareness accordingly. In one exercise, for example, we lay on top of two thick mats on our bellies, placing our hands on the floor and practising the positioning of hands, arms, shoulders and 'knitted' torsos while relieved from controlling our lower bodies. In another exercise, a special chair-shaped object with a hole in the seat area provided support for our inverted upper bodies from the shoulders. This allowed us to focus on finding the right position for our torsos, pelvises, legs and feet and to experience the sensation of handstanding without the typically required strength from the arms. In another exercise, the wall provided support: we had to walk our feet up as high as possible until we could ideally detach the feet to stand freely, thus developing awareness of the positioning of, and muscle activation around, our hands, wrists, shoulders and torsos without having to support the full weight of our bodies or be able to maintain balance on our own. We went through the circuit of exercises in pairs. While one engaged in performing the exercise, the other was meant to observe closely, give feedback on the accuracy of the movements, help bring the body into the expected posture and provide additional support. Meredith and Josh joined the pairs to supervise this mutual process by supporting the 'performer' through verbal cues and physical manipulations and by helping the 'student-teacher' scan the other's body to identify problems and find suitable instructions.

Understanding the first appearance of elephants in the room in order to understand the second

One of the benefits of an ethnographic method is the ability to trace connections between observations 'at a time' and 'over time'. The second elephant moment sketched at the outset happened in the sixth week of attending class at a station designed for learning to dissolve a handstand by cartwheeling away from the leg support provided by a vault. When we tried a similar exercise two weeks earlier, Kristina's feet had made a loud thud as they hit the ground. We laughed at the thud and said that maybe we were elephants. We both identified the thud as showing a lack of control before the landing.

The elephant moments signified a turning point in learning: we started identifying problems with our technique in the absence of the instructors pointing them out first. 'Individuals continually endeavour to improve their own performance and often draw on multiple role models of expertise', write Downey et al. (2015, 194). In the elephant moments, role models for a desirable landing included the smooth, elegant control of the instructors during their demonstration of the task, as well as some other class participants as they practised the exercise around us. Additional role models came from past experiences of alternative movement practices. The elephant experience contrasted with Kristina's memorized feel of lightness evoked by similar gymnastic movements she had accomplished with ease as a teenager. Kath was reminded of being taught by a physiotherapist to 'land silently' when jumping to absorb the impact of hitting the ground.

Drawing on multiple images and experiences, the second elephant moment shows how we were able to draw on our elephant diagnosis to guide how we approached the next attempt. Metaphorically, 'being' an elephant one week encouraged us to try to not be an elephant when landing the cartwheel move in later weeks. In the second fieldnote, we were practising the handstand-to-cartwheel move, alternating between left- and right-sided dismounts. Our aim was to land silently as a way of guiding a more elegant, controlled technique. As we joked about it, two weeks after the original diagnostic moment – 'No elephants!', 'I have an elephant on this side!', 'Mine is on this side!' – the elephants shifted from something we 'were' to something (helpful) we 'had' and could potentially shed. The exchange was one part compliment, one part acknowledgement of work still to go.

The multimodal scaffolding of movement control in the early stages of skill learning

The elephant episode exemplifies the presence and the development of learners' ability to monitor, evaluate and guide their own performance. We previously discussed the role of cues and instructional nudges in guiding complex movement sequences in experts as an example of a non-linguistic, or linguistically compressed, form of movement control. Our shared experience of the elephant episode provides insight into the modalities and constitution of such control in a novice context. There is a *bodily feel*, or kinaesthetic awareness, of heaviness at the moment of the feet hitting the floor. There is an *auditory cue* from the heavy landing at this exact moment. The cue is shared between the learners in a spontaneously, *verbally uttered metaphor* of being an elephant. This metaphor and the imagery it provoked (later: 'having' an elephant on one side, but not the other) contains an *affective element*. It is humorous, and as Kristina's fieldnote shows, it adds a sense of enjoyment to the (frustrating) experience of failure. In the first elephant occurrence, the metaphor articulated a gap identified by the two learners between the actual bodily performance and the requirements of the practice through the sound, sight (for the observer) and kinaesthetic 'feel' of landing (experienced directly for the performer and empathetically or intercorporeally (Brümmer and Alkemeyer 2017) for the observer). As such, it defined a moment of a shared, multimodal, practical reflexivity in Bourdieu's sense, with the two novices critically acknowledging that something went wrong. In the second occurrence, the metaphor functioned not only as an instance of reflecting or evaluating a movement gone wrong but as a feed-forward mechanism. The elephant playfully transitioned from something we were to something we could presumably get rid of. Trying not to be, or have, an elephant assisted us with the implementation of the movement far more effectively and holistically than paying explicit attention to each individual component of the movement. This observation aligns with studies that demonstrate that learning through analogies ('like a ... ' – or 'not like' an elephant in this case) and metaphors is a useful technique for novices in other domains as well (Brümmer 2015; Capio et al. 2020). A key feature of our case is the spontaneous, idiosyncratic development of the analogy between the two learners, rather than the analogy coming from a coach or another expert practitioner.

The socio-material arrangements of the handstands classes encouraged the development of movement control from the first lesson. It was shaped by the teachers' verbal instructions, commented demonstrations, their verbal cues and touches and the different learning stations which discriminated certain aspects of the to-be-learned practice as well as distinct aspects of the learners' bodies and attention. It was shaped by having learners work together in pairs, teaching them to closely observe their partner's body in action and having them identify problems and find instructions and feedback. It was further developed in arrangements of objects and exercises which isolated key aspects of the aimed-at handstand practice and supported learners to selectively focus their attention on selected aspects of the practice only. Our findings suggest that while movement control may transform to more efficient, non-linguistic forms with experience, as suggested by Christensen et al. (2016), some of these features of movement control may be found at very early stages of learning as well; forming early rather than (only) *trans*forming later. This indicates that it would be productive to track the development, fine-tuning and adaptation of multimodal forms of movement control over much longer timescales in a range of practitioners and practices. We suspect that multimodal strategies for movement control developed in early stages of learning foster a reliance on, and trust in, similarly efficient forms of control for scaffolding movement at all levels of skill – particularly in challenging or high-stakes situations, for error correction and in future stages of development.

Bodily obstinacy and the non-linearity of learning

The elephant episode poses a challenge to any theory that conceptualizes learning as a linear process of improvement. Experiences of failure and bodily obstinacy were not only abundant in our learning but an anticipation of failure – and the regularity of failure – was built into the classes. The cartwheeling exercise did more than enable learners to dissolve a handstand elegantly. It fostered the skills needed to land safely if we did a handstand without support and started to wobble sideways. The cartwheel move trained learners in a technique to prevent injury or uncontrolled collapse if something went wrong and accounted for the likelihood of problems at unexpected times. The exercise played a crucial role in helping us develop abilities to deal with the possibility of failure, as well as the fear of failure. Again, this affective element is important, as fear or worry can lead to tension, stress and an inhibited movement, all of which can obstruct performance (Bicknell 2010; Brümmer 2015).

The first elephant moment occurred a week after a rather successful session and further points to the importance of recognizing regression as part of the learning process. When the elephant first appeared, both of us were extremely tired. Kristina's body felt immobile and somewhat heavier than usual. She experienced her body as not obeying her will and not performing movements accomplished with much more ease the week before. On the week of the second elephant moment, it was Kath who was struggling more – her body felt 'locked up', her proprioceptive awareness was much poorer than normal. There are many (unremarkable) factors that can lead to dips, blips and regression: fatigue, stress, old and new injuries or a lack of focus, among others. Regression and imperfection are inevitable. Ongoing success requires an ability to work with failure rather than against it.

We take these observations to suggest understanding movement learning as a dynamic and unstable process. Moments of failure and regress not only abound but must be prepared for through the development of particular skillsets (in this case, techniques to dissolve a 'wobbly' handstand alongside abilities to monitor and attend to one's own body and ways of moving). Bodies cannot be manipulated and 'used' at will. Instead, physical states such as pain, injury or fatigue can obstinately get in the way of drilling and moving the body according to certain preset standards, demands and requirements. This supports and extends Bicknell's (2021) argument that an important component of skilled performance and movement control is being able to monitor, adjust and adapt to fluctuations in physiological and psychological capacities; that maintaining an awareness of the vulnerability of our bodies in precarious circumstances is essential for staying safe. Our findings also put pressure on those sociological theories of practice which claim that practices are carried and performed by 'hyperable' (and adequately socialized) bodies, without often taking into consideration, however, that these bodies are living bodies with physical weaknesses and vulnerabilities (see also Brümmer, Alkemeyer and Mitchell in press).

Conclusion

We have analysed how two novices experienced and dealt with failure and fluctuation during the learning of a movement practice. Drawing on ethnographic data, and integrating theoretical concepts from sociology and cognitive theory, we argued that learning and performing movement practices is infused with awareness, (practical) reflexivity and working with failure, not against it. Expanding claims by Christensen et al. (2016), we argued that efficient, non-linguistic forms of movement control are not only features of expert performances, but evident in novices' doings. By focusing closely on handstands classes as they actually happened, we demonstrated how this form of control may be actualized, scaffolded and distributed across the different human and material elements present in the learning context. We found movement control not only important for increasing skilfulness at performing a practice (or increasing with such skilfulness), but also as something developing through and motivated by failure. Movement control, however, does not guarantee skilled performance in any instance of learning. Our data indicate that learning is characterized by ups and downs, by progress and regress; it is a process in which inabilities and obstinacies may get in the way of the purposeful, practice-specific formation and use of bodies.

In addition, we illustrated the importance of collaboration between two novices for learning and, thus, the role of learning companions. Engaging in the handstands classes together as novices fundamentally helped our learning: modelling, supporting and observing each others' bodies enabled us to grasp the details of the aimed-at practice better; seeing the other perform the practice and being corrected by the instructors helped us attend to our own bodies in new ways; sharing negative experiences, talking about failure, articulating disappointed expectations or joking about them provided a way for us deal with these experiences and to muster the continuous motivation required for learning the complex practice. These observations challenge top-down conceptions of movement learning, which emphasize the roles of experts in training novices. Our study provides an instructive starting point for further investigating how learning and performing practices might also profit from collaboration between equally inexperienced companions – and, in some cases, their imagined, long-trunked sidekicks.[3]

Notes

1 Teachers' names have been changed for anonymity.
2 We use the term 'handstand' because this is what the class was called. Some practitioners prefer the term 'hand balancing' (Damkjaer 2018) or 'hand controlling' (Handstand & Fitness Coaching 2020). Gymnastic, acrobatic and capoeira handstands, among others, have different aesthetic and functional requirements which necessitate different forms of awareness and control (Downey 2012).

3 Kristina Brümmer would like to express her gratitude to the German Fritz-Thyssen-Stiftung for funding the research stay at the Centre for Elite Performance, Expertise and Training at Macquarie University, Sydney. Kath Bicknell's contribution to this project was funded by the Australian Research Council Discovery Project grant DP180100107 'The Cognitive Ecologies of Collaborative Embodied Skills', awarded to John Sutton (2018–20). Thank you to John Sutton, members of Macquarie University's Cognitive Ecologies Lab, participants of the work-in-progress workshops held while developing this volume and two anonymous reviewers for their insightful and enthusiastic comments during the collaborative, embodied development of this chapter.

References

Anderson, J. (1982), 'Acquisition of Cognitive Skill', *Psychological Review* 89 (4): 369–406.

Bicknell, K. (2010), 'Feeling Them Ride: Corporeal Exchange in Cross-Country Mountain Bike Racing', *About Performance* 10: 81–91.

Bicknell, K. (2011), 'Sport, Entertainment and the Live(d) Experience of Cheering', *Popular Entertainment Studies* 2 (1): 96–111.

Bicknell, K. (2021), 'Embodied Intelligence and Self-Regulation in Skilled Performance: Or, Two Anxious Moments on the Static Trapeze', *Review of Philosophy and Psychology 12:* 595–614.

Boll, T. and P. Lambrix (2019), 'Editorial: Untapped In/Abilities', *Österreichische Zeitschrift für Soziologie 44*: 261–7.

Bourdieu, P. (1990), *The Logic of Practice*. Stanford, CA: Stanford University Press.

Bourdieu, P. (2000), *Pascalian Meditations*. Stanford, CA: Polity Press.

Brümmer, K. (2015), *Mitspielfähigkeit. Sportliches Training als formative Praxis*. Bielefeld: transcript.

Brümmer, K. and T. Alkemeyer (2017), 'Practice as a Shared Accomplishment. Intercorporeal Attunement in Acrobatics', In C. Meyer and U. Von Wedelstaedt (eds), *Moving Bodies in Interaction – Interacting Bodies in Motion. Intercorporeality, Interkinesthesia, and Enaction in Sports*, 27–56, Amsterdam: John Benjamins.

Brümmer, K., T. Alkemeyer and R. Mitchell (in press), 'Building Blocks of a Historical Overview of 'Tacit Knowledge'', In A. Kraus and C. Wulf (eds), *The Palgrave Handbook of Embodiment and Learning*, London: Palgrave Macmillan.

Capio, C. M., L. Uiga, M. H. Lee and R. S. W. Masters (2020), 'Application of Analogy Learning in Softball Batting: Comparing

Novice and Intermediate Players', *Sport, Exercise, and Performance Psychology* 9 (3): 357–70.

Christensen, W., J. Sutton and D. J. F. McIlwain (2016), 'Cognition in Skilled Action: Meshed Control and the Varieties of Skill Experience', *Mind & Language* 31 (1): 37–66.

Damkjaer, C. (2018), *Homemade Academic Circus*. Winchester: Iff Books.

Downey, G. (2012), 'Balancing between Cultures: Equilibrium in Capoeira', In D. H. Lende and G. Downey (eds), *The Encultured Brain: An Introduction to Neuroanthropology*, 169–94, Cambridge, MA: MIT Press.

Downey, G., M. Dalidowicz and P. H. Mason (2015), 'Apprenticeship as Method: Embodied Learning in Ethnographic Practice', *Qualitative Research* 15 (2): 183–200.

Fitts, P. M. and I. M. Posner (1967), *Human performance*. Belmont, CA: Wadsworth.

Goodwin, C. (1994), 'Professional Vision', *American Anthropologist* 96: 606–33.

Handstand & Fitness Coaching [@coachbachmann]. (2020, December 8). Things to Know and Always Remember. Retrieved from https://www.instagram.com/p/CIg84Hog1qb/.

McIlwain, D. and J. Sutton (2014), 'Yoga from the Mat up: How Words Alight on Bodies', *Educational Philosophy and Theory* 46 (6): 655–73.

Pacherie, E. and M. Mylopoulos (2021). 'Beyond Automaticity: The Psychological Complexity of Skill', *Topoi* 40 (3): 649–62.

Pink, S. and J. Morgan (2013), 'Short-Term Ethnography: Intense Routes to Knowing', *Symbolic Interaction* 36 (3): 351–61.

Reckwitz, A. (2002), 'Toward a Theory of Social Practices. A Development in Culturalist Theorizing', *European Journal of Social Theory* 5: 243–63.

Schatzki, T. (2002), *The Site of the Social: A Philosophical Account of the Constitution of Social Life and Change*. University Park: Pennsylvania State University Press.

Sutton, J. (2007), 'Batting, Habit and Memory: The Embodied Mind and the Nature of Skill', *Sport in Society* 10 (5): 763–86.

Sutton, J. and K. Bicknell (2020), 'Embodied Experience in the Cognitive Ecologies of Skilled Performance', In E. Fridland and C. Pavese (eds), *The Routledge Handbook of Philosophy of Skill and Expertise*, 194–206, London: Routledge.

Wacquant, L. (2004), *Body & Soul: Notebooks of an Apprentice Boxer*. New York: Oxford University Press.

5

Not breathing together: The collaborative development of expert apnoea

Greg Downey

In one of my first freediving classes in Sydney, the group did a pyramiding exercise in a pool. Dressed in wet suits, six novices first held our breaths underwater while lying still for as long as possible, what freedivers call 'static apnoea'. With a partner timing us, this first attempt was the 'base' of our 'pyramid'. In the exercise, we had two minutes to recover and then submerged again, instructed to stay under until our partner, the timekeeper, tapped our shoulder. Ten seconds were added to the base, our previous 'maximum', as our new target. Although we all rolled our eyes, we also made our target times, and continued to do so even as the target was repeatedly raised. Over the next hours, we dove repeatedly, with the guidance of our instructor and the support of a 'diving buddy', and found that we achieved longer and longer breath-holds.

We had spent the first half of the day listening to a lecture and talking in a small meeting room at a dive centre, reviewing the techniques, safety procedures and science of apnoea, training to breath-hold dive with minimal equipment (diving mask, wetsuit and fins). We learned how deprivation of air affected our bodies, the physiological response to carbon dioxide build-up (or hypercapnia) and the slower onset but more dangerous effects of oxygen deficit (hypoxia). The in-pool session and pyramid exercise were especially remarkable because we achieved longer breath-holds with progressively shorter recovery periods. The pyramid's recovery times started at two minutes and decreased by fifteen seconds each attempt; our breath-hold targets often increased by ten seconds at each 'step'. This chapter explores how training, especially a regimen of self-cultivation, demonstrates collaborative enculturation and offers an analytical vocabulary for thinking about the social supports in a cognitive ecology of skill.

Not diving alone: Core theoretical concepts

On the surface, a sport like freediving might appear supremely individualistic: a person dives alone, even risking death. Breath-holding is achieved neurologically by descending inhibition of the brainstem motoneurons or the final spinal motoneurons that drive respiratory pumping (Orem and Netick 1986). But more deeply, from a holistic perspective, one that recognizes how skill acquisition requires guidance over time, the developmental foundation of skill is collaborative. Jean Lave (e.g. 1991) has inspired anthropologists of education to recognize that learning is most often a social activity, situated interactively in everyday life, a point also made by Lev Vygotsky (see Daniels 2001).

Inhibiting automatic processes like breathing is an executive capacity of an individual, refined by training and sculpted by shared psychological techniques. A diver consciously interrupts habitual tidal breathing and actively resists the impulse to breathe. My approach builds in part on criticism (such as by Christensen et al. 2016) of the popular view – defended by Dreyfus (2002), among others – that skilled action is smoothly automated.[1] Freediving is especially interesting in a discussion of the cognitive dimensions of skill because breathing is the archetypal 'automated' neurological function. During skill acquisition, the novice encounters challenges that require forms of social scaffolding and corporeal collaboration to overcome. As the Vygotskyian paradigm for sociocultural learning argues, external systemic supports become introjected and internalized during development, including in ways that seem 'automatic', but in this case, they modify the functioning of a highly 'automated' neural system.

This chapter explores stages in developing breath-holding skill to show different forms of collaboration, specifically what I call *catalysis, introjection, scaffolding* and *buttressing*. All are cooperative assemblages that facilitate practical, perceptual and neurological change, and although roughly sequential (*catalysis* being first), they are recursive and overlapping. The terms seek to highlight the many ways the individual is 'extended' in a developmental assembly, and that different 'extensions' have diverging destinies: some become obsolete, others internalized, while others must remain in place for skilful action.[2]

In regimes of self-transformation like freediving, as in all sports training, individuals manipulate their neurological development in only partially conscious ways, using social supports and interaction. Roepstorff and colleagues (2010) argue that patterned practices – behavioural forms of enculturation – shape neurological functioning through collaborative activity (see also Lende and Downey 2012). Constellations of pedagogical techniques accumulate, becoming a 'developmental niche' in which expertise reliably develops. Super and Harkness (1986) have highlighted how every child is born into a culturally specific 'developmental niche' (see also Stotz 2010). In the case of freediving, the skill's 'developmental niche' includes progressive exercises and exposure to water that transform novices, changing how they respond to immersion in water practically, psychologically and even physiologically.

The dynamic force of habit and catalysis for change

In our classroom sessions, freediving students heard that the initial impulse to breathe, usually after thirty to forty seconds, was force of habit. We were told, if we refused to breathe, our lungs might 'burn' and we would experience a growing urgency, but the discomfort was not a true signal that anything detrimental was occurring. In the scientific literature, Bloch-Salisbury and colleagues (1996, 950) refer to the 'discomfort caused by the urge to breathe' as 'air hunger'. Early research on breath-holding revealed that the sensation of being breathless arose from a complex combination of muscular and chemical perception, and that it was amenable to training effects (see Guz 1977 for a review). The ability to breath-hold varies, depending on a range of 'psychological' factors, such as a person's motivation and tolerance of discomfort (Schneider 1930). Binks and colleagues (2007), for example, highlight that the perception of 'air hunger' is dissociable from blood gas level; long-term changes in carbon dioxide levels in the blood (and blood acidification) can come to feel 'normal' to patients, such as those with chronic respiratory restrictions.

In the pool session, our instructor advised us to stay calm: arousal increased heart rate and oxygen use, making the breath-hold more difficult. We had no way to prove or disprove these claims, but our handbook had citations of scientific articles, and our instructor spoke from years of experience (and had once held a world record). Our instructor's social authority introduced a new element into the dynamic system of sensations and bodily processes that maintained habitual tidal breathing. The class was *already* destabilising our breathing pattern, *catalysing* change, so that we would attempt to hold our breaths longer than ever before. For example, in the classroom, we breathed through drinking straws to accustom ourselves to having less air available; we were told that the exercise, along with others, would increase our lung volume.

Many events might provoke a person to come to a freediving class: I had been encouraged by someone who knew the instructor. Another student reported that she learned of the course while scuba diving. Another practised breath training for sports endurance. A surfer joined the group after being dumped badly by a large wave, for personal safety and decreased anxiety. In every case, we chose to engage with the course to *disrupt our habitual breathing*, and then we looked to each other for inspiration (and to avoid shame) in the class as we pushed ourselves to stay submerged longer. From a social perspective, these initial external stimuli were *catalysts* to initiate systemic change. The catalyst caused us to assemble softly new elements into our breathing systems, including coaches, a course, a textbook and even each other as resources. These external supports might help us achieve new patterns: some became durable parts of our 'extended' respiratory systems, but others were temporary, discarded when no longer expeditious.

Coaching and introjecting self-management techniques

Given that the first obstacle to a breath-hold was habit, we set new targets, tried to downregulate our emotions and performed distractor tasks. These tasks were simple mental activities like singing a song, walking through a familiar place or doing a familiar task in our imaginations. The activities were phenomenal-behavioural manipulations, cultural techniques long established in the freediving community. Our instructor encouraged us to try a variety of distractors to see which ones we found most successful. He had his own preference but also drew on the experiences of others: when I later read about freediving, I ran across similar exercises (e.g. Pelizzari and Tovaglieri 2004, 347 f.).

Binks and colleagues (2007, 175) found expert divers invariably used distractor tasks during long apnoeas. Diving includes an education of attention in a very specific way: a capacity to distract oneself to diminish one's reactivity, both psychologically and physiologically. We were told that some of the best divers meditated or dove in a trance-like state. At this first stage of skill development, when a person increases their breath-hold beyond the habitual or 'conventional breakpoint' (Lin et al. 1974), the transformation only requires confronting discomfort and shallow self-manipulation with social and ideational support. This manipulation is probably entirely perceptual in scale, not affecting the nervous system or physiological responsiveness in significant ways. It happens quickly. But it affects behaviour, expands activity, rearranges the diver's relation to the environment and brings the diver's body through activity to the doorstep of the next breath-hold event or major threshold in expertise.

These instructions resemble many other explicit forms of coaching and formulae that students are told: how to 'duck dive' to quickly convert lateral momentum to descend in the water, how to 'hook breathe' when surfacing to quickly recover from the breath-hold without panting, how to signal we are okay after a dive so we are not penalized in competition. Some of these coaching techniques are 'instructional nudges' (Sutton 2007, 773), densely meaningful mantras that cue a chain of actions; others are sequences that can be automatized or done largely by rote. Through these 'nudges', crucial coaching techniques are taken on board and internalized as expertise grows, transforming social support into self-monitoring, so that the subject can extend and stabilize his or her own performance. Some of these lessons must be tested against the idiosyncrasy of the person's own preferences and responses. For instance, several techniques are available to equalize the pressure between the water and one's ears to prevent them from being damaged as a diver goes deeper. Which technique works depends on the diver and depth, and the best approach might change over time, so knowing several is recommended; some strong freedivers find equalization is the most persistent challenge.

This type of collaboration – what I am here calling *introjection* – is well discussed in the sociocultural learning theory inspired by Vygotsky: his approach highlights that many individual capacities are first experienced through collaboration, as a more experienced teacher supports the novice. As we attempted longer static apnoeas, our coach shared other techniques and tried to give tailored guidance, sharing his experience and supporting our exploration of our own capabilities.

Scaffolding and desensitising respiratory interoception

Before our pyramid exercise, we were told that, if we continued to hold our breath, probably around a minute or minute and a half later, our diaphragms would clench painfully. Divers call this a 'contraction'. Our instructor told us it resulted from carbon dioxide build up and blood acidification. We could safely ignore it, he insisted. After the first contraction, we would briefly feel more comfortable, before a second, and then at increasingly short intervals, more contractions. I felt my first contraction as my target times grew longer. As they passed, they became less distressing.

Researchers studying breath-holding have found that, if a person continues to hold his or her breath as carbon dioxide accumulates, the diaphragm muscles will begin involuntary movements (Agostoni 1963). These involuntary activations of the inspiratory muscles were called, by Lin and colleagues (1974), 'involuntary ventilatory activity' or IVAs (there are also sometimes called 'involuntary breathing movements' or IBMs; e.g. Breskovic et al. 2012). Lin and his colleagues (1974) argued this was the 'physiological break point' of a voluntary apnoea for a 'naïve diver'. We were becoming non-naïve; instead, we were told to disregard and count them, to see how many we could endure.

After IVA begins, divers enter the 'struggle' phase of the dive (Lin et al. 1974). In the struggle phase, an apnoeist will experience repeated 'contractions', growing in intensity and frequency, according to researchers, until they may be intolerable. That said, Bain and colleagues (2018) report that experts can experience seventy-five or more IBMs before terminating a breath-hold. Empirical research (2018) shows that for the expert diver, severe hypoxia, not CO_2 build-up, forms the break point, and some divers approach 50 per cent blood oxygen levels – the theoretical minimum before blackout.

To tolerate contractions, blunt their effect and desensitize oneself to them involves perceptual learning, a change to one's sensory acuity brought about by routinization and repeated exposures. Without collaboration and the *scaffolding* of an experienced instructor, the naïve diver will likely never realize that increased expertise lies beyond this claxon of bodily sensation. In contrast to halting habitual breathing rhythms, this phenomenal-behavioural manipulation likely does have measurable effects on nervous system responsivity, potentially even phenotypic adaptation (Ostrowski et al. 2012). Bain and colleagues (2018, 645–6) highlight that experimental results are contradictory: whether the chemosensitivity of veteran divers is decreased or they simply tolerate higher levels of response (more contractions) is not clear.

With the scaffolding of information about IVA and a coach's encouragement, divers like my class experiment with disregarding the 'physiological breakpoint'. In the process, experts either blunt their sensitivity to hypercapnia or better tolerate IVA, emotionally and practically. Like experts, after we had our first IVA, my class recovered and went under again, with knowledge that led us to reinterpret our bodily signals, try to override our impulses, and hold our breaths longer than our previous maxima. We surfaced, took progressively shorter recoveries and increased our targets. Over time, we would develop our own familiarity, and the necessity of having a coach to soothe and encourage us would diminish: like *scaffolding*, this support could be removed.

Buttressing and phenotypic adaptation

Despite the discomfort of contractions during our pyramid exercise, we were advised that hypercapnia could not harm us. Hypoxia, or oxygen deficit, would hurt us before carbon dioxide surplus. Hypoxia became a threat after four or five minutes, as less oxygen arrived in the brain. That day, none of us held our breath so long that hypoxia became a threat and we risked unconsciousness. Some medical researchers on breath-holding assumed that a person could not voluntarily hold their breath until loss of consciousness (e.g. Parkes 2006), but, in fact, competitors regularly induce unconsciousness in competition. This danger highlights the fourth collaborative relationship that supports diving expertise: *buttressing*.

Buttressing is a form of 'extension' that remains in place as a contingency; in freediving, some buttressing is done with equipment. Competitions, especially the depth disciplines, include dive lines that measure depth, but also act as a safety rope; every competitor dives while connected to it by lanyard. Proper safety equipment, including buoys to prevent divers from being run over by boats, were introduced in our earliest lessons because of the necessity of material buttressing for breath-holds. But divers also use a dive 'buddy' to observe them; we were told we should never dive alone. The reason was that, the better we became at resisting IVA, the more dangerous diving became. Veteran divers claim to experience a decrease in 'struggle' as apnoea lengthens; IVA decreases. According to Binks and colleagues (2007, 175), the expert freedivers they interviewed all reported using cognitive self-checks in long breath-holds as they risk blackout (at six to nine minutes). They did mental arithmetic or checked to see if their colour vision still functioned to alert themselves if they risked falling unconscious. These divers insisted that their 'air hunger' subsided, and their fear of losing consciousness, and thus being disqualified in competition, caused them to self-monitor closely.[3]

Divers submerge themselves into an environmental niche – underwater – where repeated exposure will diminish their responsiveness to hypercapnia, but they risk overshooting expertise to put themselves in danger. At the same time, repeated and prolonged immersion provokes and reinforces the 'dive reflex,' an evolutionarily ancient response in all vertebrates (Foster and Sheel 2005; Schagatay et al. 2000). When a person's face and nostrils detect water and apnoea occurs, the human dive reflex is provoked, and the vagus nerve triggers bradycardia, or a slowing of the heart rate; this is why splashing cold water on the face can slow heart rate when a person is nervous. The same neurological reflex is especially pronounced in aquatic mammals like seals, otters and walruses, but it is conserved across air-breathing vertebrates and has homologies in fish. The reflex occurs automatically when an air-breathing animal is submerged, especially strongly when cold water contacts the face and nostrils; the reflex overrides basic homeostatic reflexes and stress to use oxygen more efficiently.

If the apnoea is sustained, the dive reflex becomes more pronounced, and the body undergoes a sympathetically induced vasoconstriction, a narrowing of peripheral arteries which shunts blood to core organs (Heistad et al. 1968). Longer still, and the spleen may contract, dumping red blood cells to increase oxygen in the bloodstream. In the most extreme cases, cardiac arrhythmia or irregularities occur. In diving mammals, these responses are dramatic; in humans, they are trainable. Initial responses grow more rapid and profound in veteran freedivers (Schagatay et al. 2000). The heart rates of elite apnoeists, for example, can drop to twenty to thirty beats per minute, some of the lowest on record, as they intentionally downregulate stress. Some meditate, but community debate continues about the best way to prime the nervous system.

A dive reflex and severe bradycardia is the sum of many resources: vertebrate evolution, deeply conserved neurological programmes, emotional self-regulation, bodily priming through cultural training, community learning from both historical practice and science and even escalation of performance and competitions. But the training required to fully manifest the dive reflex is arduous and risky, so divers require collaborative *buttressing* to pursue adaptation in their nervous systems.

Integrating cognitive ecology and embodied culture

In advocating for 'cognitive ecology' as a conceptual approach, Tribble and Sutton call for a focus on the 'rich ongoing interaction with our environments' (2011, 94) to bring together insights on social and distributed cognition. As they write, 'mental activities spread or smear across the boundaries of skull and skin to include parts of the social and material world' (2011, 94). Expert apnoeists demonstrate that executive function, bodily self-control, even downregulation of highly automatized functions can become thoroughly social and embedded with the material world. In the case of breath-holding, the complexity of interaction is partially within the divers themselves, as ideation, techniques, even competition rules, some obviously 'cognitive' and others less so (or even nonconscious), influence the nervous system and organs, such as the lungs and diaphragm, especially with sustained training. The assembled system that produces skill also includes the community practices and social stimuli (like competition), as well as the diving equipment assembled for the activity: wetsuits, ballast, masks and the distinctive long fins worn by freedivers. Hutchins has suggested that 'activity in the nervous system is linked to high-level cognitive processes by way of embodied interaction with culturally organized material and social worlds' (2010, 712). The apnoeist is a developing soft assemblage, an individual embedded in and engaged with a network of social relations and material resources, some of which are developmental and transitory like *catalysts* and *scaffolding*, others of which remain necessary *buttressing* or become *introjected*. Expert breath-holding is social at its foundation.

Breath-holding also depends on an aquatic milieu which may be experienced as foreign, buoyantly supportive but also dangerous. The material properties of water become a constituent partner in extending the breath-hold. Immersion proves that the athlete has maintained the apnoea, but it also triggers the distinctive dive reflex, some parts of which are dormant in most people, only unmasked and made manifest by repeated exposure to water. In freediving, the diver (and coach) partner with water itself to provoke a physiological reaction. The neurological responses are built up and leveraged into phenotypic change by systematic exposure to air hunger in water. Although it might be possible to stumble upon this developmental potential without outside guidance, the experience of veteran divers, their knowledge of how the nervous system interacts with water, even scientific research on apnoea and the dive reflex, provides the novice with guidance. The path to expertise is signposted by the community.

The day in the pool, on the last breath-hold of the pyramid, after our shortest recovery of thirty seconds, I achieved a personal best: over two minutes for a breath-hold. I felt four contractions in the 'struggle' phase of my apnoea, but they felt less violent than the first. All the students experienced first-hand how we adapted; our response to CO_2 build-up felt blunted and easier to manage.[4] As we reinterpreted our own bodily sensations and controlled our breathing, we expanded control over our bodily systems on the way to greater ability.

Conclusion: Skill acquisition as developmental assemblage

Skills are most often built through collaboration, even where the resulting activity appears solo, like freediving. Freediving demonstrates the layered way practices make use of physiological potential and developmental processes, and the sophisticated role cognition plays, including indirect contributions, like doing apparently unrelated distraction exercises that help skilful activity. Skills are not simply the 'sedimenting' of naïve experience or the gaining of 'know how' but also the building of 'so that' in the nervous system: the canalization of the body's plasticity and physiological development to produce new capabilities. Case studies like a neuroanthropological analysis of freediving help us to understand cultural patterns of human variation more broadly and produce more accurate accounts of cultural embodiment. In anthropology, we sometimes find accounts of enculturation that treat skill acquisition as a 'transfer' of something ineffable between expert and novice, an 'information' transmission. The socially constructed developmental assemblage provides a more elaborate and guided incubator for skill, including physiological and neurological adaptation (Ostrowski et al. 2012). A coach cannot 'transfer' a blunted CO_2 reflex to a novice but rather creates a *scaffolded*, *buttressed* environment where a student can work on their own psychological and physiological responses, providing advice that might be *introjected* by the student to aid self-control.

Breath-holding demonstrates dramatically how individual cognition can be used, with social supports, cultural insights and the environment, to expand a diver's physiological capabilities. This capacity – to create practical assemblies to sustain training with developmental consequences – is social, not just because imitation occurs, but also for a host of other reasons. First, the training regime emerges from a community of practice, based upon the accumulated experience and insight of practitioners who have learned this skill before the novice, and their goals, rituals and preferences. The individual who gets in the pool to do a 'pyramid' confronts the exercise as an established practice or traditional activity, but it is the product of shared, collaborative learning, experimentation and observation. Years of training, thousands of practitioners, competitions and world records all contribute. Countless hours in pools and oceans – experienced, observed and interpreted, sometimes accurately and sometimes not – bequeath to the novice a shared 'wisdom' to which the individual subjects him- or herself. Over time, these socially derived techniques for bodily self-cultivation change. Individuals fail at, evade or innovate with them. Records are broken. New equipment is invented. Practices once considered 'essential' fall into disuse, replaced by new pedagogies or techniques. Activities, once accepted, like 'no limits' diving (using equipment that took divers to over 200 metres depth and helped them ascend rapidly), might be judged 'too dangerous'. When an individual seeks to learn a skill, they confront this social avenue to expertise. Sometimes, autodidacts or splinter groups invent their own paths, like the iconoclasts who have innovated in freediving. But for most novices seeking to hold their breaths longer, a community directs them to try a pyramid exercise in a pool.

The case of breath-hold diving suggests multiple reasons to talk about 'cognitive ecology' in a neuroanthropological study of skill acquisition, including: 1) recognizing that cognition is 'embodied', not just that it depends on biological *components*, but also that it has neurological *consequences*, especially in areas like skill acquisition with sustained training; 2) realizing that skill acquisition and 'cognition' include placing the developing nervous system into particular relations with the environment, equipment and other people, as both facilitator and challenge, in constellations structured by cultural knowledge; and 3) noting how the developmental niche or assembly is progressively reconfigured; these relations are changing developmental assemblages as skill accretes, capacities increase and perceptions become more acute. The coach must know when to step back, when to take the novices out of the classroom and put them in the pool so they can go to work on their own perceptions and reactions, helped by buddies and water.

Exercises like the pyramid breath-hold are sophisticated sensory and behavioural interventions individuals undertake on their own bodies, with the collaborative guidance and practical support of others. The training exercise is not the skill novices wish to develop; most do not seek to lie still in a pool. Rather, they want to skin-dive, spearfish, freedive in the ocean, or prepare themselves to be dunked badly when surfing. Or they dive to pursue recreational, physical, psychotherapeutic, even contemplative or spiritual goals. The pyramid breath-hold is a social technique for sculpting themselves, skill-building with phenomenological and physiological consequences, assembled out of heterogeneous parts.

This chapter sketches different categories of constitutive parts in the skill acquisition assemblage: *catalysis*, *scaffolding*, *introjection* and *buttressing*. I have discussed the cultural and cognitive implications of the intentional body work which undergirds many skills, how it blends what are typically considered 'cognitive' activities with physical adaption, sensory change and social interaction. Skills require higher order strategic dimensions, but they also involve hard-won physiological, perceptual and executive transformations of the individual, usually built in cooperative developmental assemblages. Skill is a way we use social and material forces to transform ourselves.[5]

Notes

1 In anthropology, a parallel debate has asked whether habitual actions are necessarily non-conscious (Downey 2010).
2 The concept of 'extension' here draws on the discussion of 'extended cognition' inspired by Clark and Chalmers (1998).
3 As John Sutton pointed out, this 'danger' highlights the degree to which the 'ecosystem' of freediving competitions has shaped this community's practice and knowledge. Freedivers in competitions are motivated by structured incentives to dive in peculiar ways, creating quite different skills, for example, from traditional divers who foraged or harvested pearls (see Ferretti and Costa 2003).
4 Ironically, my instructor later revealed to me that I likely had not experienced 'real' hypercapnia blunting: that adaptation required

longer-term training. Instead, I was having a purely 'psychological' decrease in my distress and reaction.
5 My gratitude especially to Jesús Ilundáin-Aguruzza, John Sutton, Kath Bicknell, Ian Maxwell and Sara Kim Hjortborg for their generous and constructive feedback on this piece, and to Daniel Lende, who has shaped the neuroanthropological approach I use.

References

Agostoni, E. (1963), 'Diaphragm Activity during Breath Holding: Factors Related to Its Onset', *Journal of Applied Physiology 18* (1): 30–6.

Bain, A.R., I. Drvis, Z. Dujic, D.B. MacLeod and P.N. Ainslie (2018), 'Physiology of Static Breath Holding in Elite Apneists', *Experimental Physiology 103*: 635–51.

Binks, A.P., A. Vovk, M. Ferrigno and R.B. Banzett (2007), 'The Air Hunger Response of Four Elite Breath-Hold Divers', *Respiratory Physiology & Neurobiology 159*: 171–7.

Bloch-Salisbury, E., S.A. Shea, R. Brown, K. Evans and R.B. Banzett (1996), 'Air Hunger Induced by Acute Increase in PCO_2 Adapts to Chronic Elevation of PCO_2 in Ventilated Humans', *Journal of Applied Physiology 81* (2): 949–56.

Breskovic, T., M. Lojpur, P.Z. Maslov, T. J. Cross, J. Kraljevic, M. Ljubkovic, J. Marinovic, V. Ivancev, B. D. Johnson and Z. Dujic (2012), 'The Influence of Varying Inspired Fractions of O_2 and CO_2 on the Development of Involuntary Breathing Movements during Maximal Apnoea', *Respiratory Physiology & Neurobiology 181*: 228–33.

Christensen, W., J. Sutton and D. J. F. McIlwain (2016), 'Cognition in Skilled Action: Meshed Control and the Varieties of Skill Experience', *Mind & Language 31* (1): 37–66.

Clark, A. and D. Chalmers (1998), 'The Extended Mind', *Analysis 58* (1): 7–19.

Daniels, H. (2001), *Vygotsky and Pedagogy*, London: Routledge.

Downey, G. (2010), '"Practice Without Theory": A Neuroanthropological Perspective on Embodied Learning', *Journal of the Royal Anthropological Institute 16*: S22–S40.

Dreyfus, H. L. (2002), 'Intelligence without Representation – Merleau-Ponty's Critique of Mental Representation: The Relevance of Phenomenology to Scientific Explanation', *Phenomenology and the Cognitive Sciences 1* (4): 367–83.

Ferretti, G. and M. Costa. (2003), 'Diversity in and Adaptation to Breath-Hold Diving in Humans', *Comparative Biochemistry and Physiology Part A 136*: 205–13.

Foster, G. E., and A.W. Sheel (2005), 'The Human Dive Response, Its Function, and Its Control', *Scandinavian Journal of Medicine & Science in Sports* 15: 3–12.
Guz, A. (1977), 'Respiratory Sensations in Man', *British Medical Bulletin* 33 (2): 175–7.
Heistad, D. D., F. M. Abbound, and J. W. Eckstein (1968), 'Vasoconstrictor Response to Simulated Diving in Man', *Journal of Applied Physiology* 25 (5): 542–9.
Hutchins, E. (2010), 'Cognitive Ecology', *Topics in Cognitive Science* 2: 705–15.
Lave, J. (1991), 'Situating Learning in Communities of Practice', In L. B. Resnick, J. M. Levine, and S. D. Teasley (eds), *Perspectives on Socially Shared Cognition*, 63–82, Washington, DC: American Psychological Association.
Lende, D. H., and G. Downey, (Eds.) (2012), *The Encultured Brain: An Introduction to Neuroanthropology*. Cambridge, MA: MIT Press.
Lin, Y. C., D. A. Lally, T. O. Moore, and S. K. Hong. (1974), 'Physiological and Conventional Breath-hold Breaking Points', *Journal of Applied Physiology* 37 (3): 291–6.
Orem, J., and A. Netick. (1986), 'Behavioral Control of Breathing in the Cat', *Brain Research* 366 (1–2): 238–53.
Ostrowski, A., M. Strzała, A. Stanula, M. Juszkiewicz, W. Pilch, and A. Maszczyk. (2012), 'The Role of Training in the Development of Adaptive Mechanisms in Freedivers', *Journal of Human Kinetics* 32 (1): 197–210.
Parkes, M. J. (2006), 'Breath-holding and Its Breakpoint', *Experimental Physiology* 91 (1): 1–15.
Pelizzari, U., and S. Tovaglieri (2004), *Manual of Freediving: Underwater on a Single Breath*. Reddick, FL: Idelson-Gnocchi Ltd.
Roepstorff, A., J. Niewhöner, and S. Beck (2010), 'Enculturing Brains through Patterned Practices', *Neural Networks* 23: 1051–9.
Schagatay, E., M. Van Kampen, S. Emanuelsson, and B. Holm. (2000), 'Effects of Physical Apnoea Training on Apneic Time and the Diving Response in Humans', *European Journal of Applied Physiology* 82: 161–9.
Schneider, E. C. (1930), 'Observation on Holding the Breath', *American Journal of Physiology* 94: 464–70.
Stotz, K. (2010), 'Human Nature and Cognitive-Developmental Niche Construction', *Phenomenology and Cognitive Science* 9: 483–501.
Super, C. M., and S. Harkness (1986), 'The Developmental Niche: A Conceptualization at the Interface of Child and Culture', *International Journal of Behavioral Development* 9: 545–69.

Sutton, J. (2007), 'Batting, Habit and Memory: The Embodied Mind and the Nature of Skill', *Sport in Society 10* (5): 763–86.
Tribble, E., and J. Sutton (2011), 'Cognitive Ecology as a Framework for Shakespearean Studies', *Shakespeare Studies 39*: 94–103.

6

Cultivating one's skills through the experienced other in aikido

Susanne Ravn

The Japanese martial art practices of Aikido are based on the ideal of winning the fight without harming the attacker (Palmer 2002; Westbrook and Ratti 1970/2001). To quote sensei Victor Merea, with whom I have trained for about eighteen years, aikido is 'to win the fight without fight'. In concrete terms, aikido is a partner-based practice where each of the two practitioners takes turns being the person attacking (*uke*) and the person exercising one of the techniques (*tori*).[1] The grips, blows or punches performed by *uke* are preset in most of the practice, following instructions and presentation of the sensei in charge of the session. In this special kind of antagonistic interaction, *tori* strives to 'harmonize' through 'blending' with the energy of *uke*, and to be the one controlling the situation by redirecting the energy of *uke* to throw or pin him/her (Kohn 2003; Palmer 2002). The practice also involves the training of falling (performing '*ukemi*') so that, when attacking as the *uke*, you can take good care of yourself. To be a good *uke*, you are expected to develop a certain flexibility and ability to attune your movement, so you can follow the redirection of your attacking movement (as this redirection is handled by *tori* exercising the technique) and meet the floor in a relaxed yet dynamic way. In the terminology of aikido, you should be able to harmonize with the energy of *tori* as well as of the fall. As I know from experience, if, in the role of *uke*, you do not harmonize – that is, if you move too stiffly, with awkward timing or try to (over)resist the technique exercised against you – *tori*'s grip and movements will make your wrist, elbow or shoulder hurt. If you resist the falling, your hip and backside will become sore and bruised.

Anchored in my own experiences and drawing on ethnographic methods, in this chapter I investigate the ecology of practising aikido. As indicated in my brief introduction to the martial art form, the practice is centred on the antagonistic interactions of two people moving together as well as aikido ideals of harmonizing with and within this antagonistic setting. That is, practitioners seek to harmonize while attacking as well as to harmonize while taking control over and pacifying the attacker. In the analysis, I aim to describe and understand practitioners' intentional engagement in this complex antagonistic interaction. I do so by recognizing that the interaction between the two practising is enabled and constrained not only by specific techniques but also by codes of conduct, ideals and connections to the Japanese heritage of martial art practices. Accordingly, I first focus on describing what enables and constrains physical practices and awareness within the aikido ecology. I describe how the enabling and constraining aspects are linked and part of a network of apprenticeship learning connected to a Japanese heritage of martial art practices. I then focus on the interactional dynamics between the two practising together by zooming in on a specific situation where I practised with a highly graduated Japanese practitioner and, in a metaphorical sense, was 'eaten up' in the interaction. My execution of the technique, and in that sense 'my control', was 'taken over' by the other. Theoretically, the analysis is informed by recent discussions connecting ecological psychological and enactive, phenomenological approaches (e.g. Baggs and Chemero 2021; McGann 2014, 2020).

As indicated in the introduction, my practice has been closely connected to sensei Victor Merea since I began, and, through this, to the relatively dynamic style of aikido aikikai, which is trained at the Reishin Dojo in Odense, Denmark.[2] I have, when possible, trained in aikido aikikai dojos in additional locations including Melbourne, Exeter (UK), Paris and in the Hombu Dojo in Tokyo – the 'headquarters' of the aikido federation. The analysis specifically draws on my participation in a two-week training period at the Hombu Dojo and observational notes from the Reishin Dojo. The latter observations were generated over a two-month period during which I used a surgery-induced break as an opportunity to observe regular practice while sitting at the side watching. These events took place in 2012–13. Several circumstances over the intervening years have delayed analysis. One important and challenging circumstance has been to find the constructive approach to handle my auto-ethnographic involvement. The oxymoronic characteristics of ethnographic fieldwork, where one strives both to connect with the participatory knowing of a practice and to create a reflective distance to these processes at the same time, have been extra challenging. I started practising aikido because I aimed at evolving my skills and competences of moving and interacting with a partner and had to find a way to create a reflective distance from a deeply incorporated knowledge I had engaged in for personal reasons. The many years that have passed since generating my notes and working out the first versions of the analysis have facilitated my ability to handle descriptions of my own experiences from 2012–13 as a special kind of data (Ravn and Hansen 2013). The temporal distance in years to the data has helped me develop a constructive ethnographical 'distance' (Hammersley and Atkinson 2007, 90) to the already incorporated skills and understandings.

Becoming part of the aikido ecology

When I originally started aikido, clear instructions set out codes of conduct I would be expected to follow when practising. Like other newcomers at Reishin Dojo, I was kindly instructed in how to enter the dojo, when to bow, where and when to sit in *seiza* (on your knees) and how to show appropriate respect for more experienced aikido practitioners. After very few visits I was expected to follow the aikido 'dress code', wearing a white *gi* (uniform), the appropriate belt colour (white for beginners) and, later on, after having graduated a couple of times, a *hakama*, the traditional black Japanese trousers. Following these codes of conduct and wearing the *gi* and the *hakama* provides access to practice, a sense of interconnectedness with other aikido practitioners, Japanese martial art traditions and grand stories of aikido.

On the one hand, aikido presents a relatively new Japanese martial art form, on the other hand, the ideals of harmonizing draw on eastern martial art traditions spanning centuries. Aikido was founded by Morihei Ueshiba (1883–1969), also referred to as *O'Sensei* (the great master). As emphasized in aikido books, articles and in the actual practice, aikido was developed out of *O'Sensei*'s spiritual and martial training, which integrated judo, sword, spear and the hard jujutsu practice with principles of Shinto and Zen Buddhism during the 1930s (Kohn 2003; Westbrook and Ratti 1973/2001). The Japanese aikido federation was founded in 1948. Taken together, these traditions and ideals continue to shape the practice of how one moves in centred ways, what it means to be aware of the energy of the partner and, not least, how to show respect for other aikido practitioners. From the day newcomers enter the dojo, they are part of a complex apprenticeship network that is spelled out in accordance with a Japanese-founded hierarchy involving the years trained and the grading examinations one has passed. In accordance with Japanese apprenticeship tradition, the sensei of the dojo will often see themselves in an apprenticeship relation to other more highly ranked senseis. In preference to spotting talent and appreciating achievements in performances, as often characterizes sports contexts, aikido practitioners are appreciated for their devotion and stamina (Kohn 2003).

When sensei presents a technique to be trained in pairs, everyone silently watches sitting in *seiza* and Japanese terms are used to describe the technique exercised (Figure 6). They are then expected to respectfully bow to their partner before and after practising a given technique, and to take turns in acting as *uke* or *tori* after four attacks. These codes of conduct 'in action' present a performative aspect of the practice and express respect and devotion to spiritually based ideals about harmonizing. However, they are more than that.

Firstly, these aspects of the aikido practice both enable and constrain the concrete physical practices and partly constitute an aikido ecology. Secondly, wearing the *gi* and *hakama* – my bodily sensation of the relatively stiff cotton material, the way the pants and the *hakama* are sewn and the feel of the relatively stiff and broader backside of the belt of my *hakama* supporting my lower back – have become part of a certain bodily sensation. The sensations of

FIGURE 6 *Sensei practising the technique* kotagaishi *with Ito-san in Reishin Dojo. Photo by famenext (@famenext).*

my *gi*, belt, *hakama* and my bare feet on the aikido mats interweave with the features of the bodily style I use, and have become, over many years of practising. They are closely connected to the way I sit in *seiza*, respectfully bow and take turns. My bodily sensations are dynamically interwoven with codes of conduct and thereby, in subtle ways, the ecology of aikido. This dynamic attunement exemplifies how qualities of practitioners' attention and sensory awareness are cultivated alongside their physical skills and abilities (Mingon and Sutton 2021) and quite actively framed by the Japanese heritage of the practice.

Scaffolding the practices of the less experienced

Shomen uchi ikkyo was the first technique I learned. *Shomen uchi* denotes the kind of attack *uke* will use when raising her arm to hit the very top of *tori*'s forehead. As *tori* performing the *ikkyo* technique, you will meet *uke*'s hitting arm at the elbow and the wrist, respectively. You will, however, focus on the elbow. Meeting the energy of the attack here will give you an entrance to unbalance *uke*. The technique demands a strong sense of timing – not just in relation to the hitting arm, but specifically in relation to how your hip initiates the movement of your upper body and your arms, and how you place yourself in relation to *uke*'s movement.[3] The movement patterns you are to perform are in a way quite simple: arms in a circle initiated from your hip while moving forward to meet *uke*'s attacking arm while taking three steps (e.g. Ueshiba 2005, 82–3). In the beginning phases of learning the technique two decades back, I recall how I intensively watched out to get a hold on the arm of *uke* attacking. When practising the technique today, I realize that I do not watch the arm as such. I first of all 'see' *uke*'s body moving: I am highly aware of sensing the tension and dynamic of *uke*'s movements just before lifting their attacking arm, and I am specifically aware how their movements are connected in their body. I realize these sensations by recalling the practice. I also recall my experiences of how challenging the technique can be if *uke* is taller or heavier than myself and how I then need to adjust to *uke*'s size and movement – the direction and angle of how I meet their elbow. Thus, not only the energy, but also the size, weight and flexibility of *uke* invites me to explore the ways this technique can be enacted anew.

I learned the *ikkyo* technique while observantly engaged in incorporating the aikido codes of conduct. Interested in incorporating the indigenous logic of the aikido practice, I strived to take care that I both performed a certain pattern of movement correctly and enacted my manners in 'the aikido way'. I was engaged in imitating shapes and patterns of movement and codes of conduct all at once. At that time, when I thought I had incorporated the *ikkyo* technique and thought I knew the patterns and logic, Sensei continued to show me still more fine-tuned aspects of the technique. While practising, he would let me know by showing what could be performed still better or with variation. At times, he would take on the role of *uke* or *tori* himself to demonstrate, through the interaction with me, where the weaknesses lay in my version of exercising the technique. He invited me to sense differences and possibilities of the technique through my interaction with him. Countless are the times he has reminded me of small details and shown me yet new facets of a technique I thought I knew so well. While enacting a technique in interaction with him as the experienced other, he invited me to be aware of specific aspects of my timing, the connection to my hip movement and the angles in use when meeting the energy of his attack. In my ongoing apprenticeship interaction with Sensei, observation, action and kinaesthetic feelings came – and still come – together while cultivating my embodied aikido insights. Practising skills which I have already incorporated, I add enhanced sensation, awareness and reflections to my 'reflex'-like enactment of the technique: I 'bring memory and movement together' (Sutton et al. 2011).

In his analysis of the apprenticeship learning taking place between the capoeira master and the capoeira practitioner, anthropologist Greg Downey has highlighted how 'the instructor's assistance helps to control the learner's body, allowing the student to execute actions that will eventually flow with much less effort' (2008, 207). The master thereby offers scaffolding assistance. Downey (2008) further emphasizes that this scaffolding process need not be part of a planned learning strategy but can run in an ad hoc way taking shape in the interaction. Downey's descriptions easily transfer to practising aikido. Depending on the levels of expertise and energy of the practitioners, Sensei adjusts the interaction he offers and the way he invites practitioners to tap into the tacit knowledge of his experienced aikido body. Accordingly, the practitioner's learning is contingent on being with and accepting the timing and energy as it comes to live in the interaction.

At Hombu Dojo, practising with one of the 'crocodiles'

When I read my notes from the two weeks practising in Hombu Dojo, a large part concerns the ongoing networking I was expected to engage in when visiting. It looks random when practitioners pair up, by bowing, to another practitioner at the very beginning of the one-hour session. It is not. As a visitor you must find ways to make contact and create an appointment for practising a day ahead or arrive early in the morning so there is time and opportunity to look out for a partner. Experienced partners are highly valued as they will offer you great possibilities to establish a good flow in the interaction when practising the different techniques presented by the sensei in charge of the session. Many notes concern this experience of being caught up in the flow of practising with good partners. Several times, I have noted how I felt I shifted my way of 'seeing' the movement of the other aikido practitioner – the experienced other. I aimed to see everything at once and deliberately used what, to my experience, was my peripheral vision to 'see' and sense the kinetic flow of my partner's movement. At that time, about eight years ago, I started to become aware that I held too much tension in my lower back, as if the centre of my movement had moved (too far) towards my lower back. When practising with different aikido practitioners in Hombu Dojo, I specifically aimed at relaxing that tension and finding a way to feel (and be) centred anew when meeting and blending with the energy of the movement of the other practitioner. Despite, or alongside, being attuned towards harmonizing the interaction with the experienced other, I also deliberately worked on attuning to what I experienced as belonging to an inner sensation of my body. I intentionally juggled my sensorial awareness to cultivate my aikido practice.

Another Danish sensei visited the Hombu Dojo at the same time as our small group from Reishin Dojo. One afternoon, he let me know with a knowledgeable smile that the highly experienced Japanese practitioners, whom I had noticed gathered in the corner of the dojo next to the balcony every morning, were also known as 'the crocodiles'. They had practised daily for decades at Hombu Dojo. Arriving early for the morning practices, they watched the new visitors entering the dojo to practice. Like crocodiles, they would silently look out for practitioners whom they, in a metaphorical sense, could 'eat up' during the impending practice. Blending with the energy and timing of your movements, 'they give you no chance to escape' as the Danish sensei explained to me. Setting up an unrelenting pace, they will wear you down. However, being invited to practice with one of these crocodiles is, at the same time, an honour – and an implicit credit to you as an aikido practitioner. They are the highly experienced others you should strive to practice with. Taka-sensei was one of these crocodiles and a close contact of my Sensei.[4] Sensei might have talked about and pointed out the small group of practitioners visiting from his dojo. At that time, I was already the oldest and most experienced of his students and I am quite sure that the communication with Sensei directed Taka-sensei's 'crocodile attention' towards me.

One early morning at the Hombu Dojo, Taka-sensei invited me to practice with him. At the time, Taka-sensei had practised at the dojo for more than forty years. I estimated that he was close to seventy, so – crocodile or not – I could not help first of all perceiving him as an old man. Taka-sensei did not mark his higher aikido status directly. Nor did he act like a master teaching. He said nothing but just started practising the technique in an ongoing flow. I remember that as the very first thing I felt, quite strongly, that I had to prove that I had the capability to blend with the energy of this flow. As I had been warned, in less than ten minutes I felt he was 'eating' me up – slowly and in full control. He was both with me and continuously ahead of me. Before I had raised my arm to perform the attack, he was already in the process of moving my movements. His movements seemed minimal – small steps without stillness – while I sensed that he had full control of my centre and that he could throw me at any moment. After a while, I tried, in moments, to also focus my attention on his arm movement to better learn from what he was doing. Somehow, he moved his arms in a stiff and strangely imprecise way compared to the precision I had aimed at developing for decades. However, the movements nevertheless felt strong, connected to his centre and extremely effective. He knew exactly where my balance was and anticipated my movements, manipulating my moving body with minimal force.

As always, the roles of practising as *tori* or *uke* changed after the four attacks. After having exercised the technique, as *tori*, he followed me as my *uke*. I immediately sensed a thick resistance in his movement and realized that I had to really push (too much according to the aikido ideals) to make him move and that the tension in my body rose. He moved, but he also let me know in the way he followed the technique I exercised that I was not the one in full control. There was nothing else to do but to try to find a way to let the tension dissolve and to fine-tune ways to let the circles of the technique begin from my hip movement, relaxing my arm movements still more while breathing, grounding, focusing on the sense of my centre. The more I found a way to relax and breathe and stay in a flow of exercising the technique, the better he moved with me. A few times, he set the tempo of the ongoing flow a bit slower and gave me a chance to be aware of particular aspects of the technique. I watched his movement – in these slow parts – intensively without really being sure if I got what he aimed at directing my attention towards. He definitely offered a scaffolded opportunity, but in a quite open way. I had to tap in and figure things out by aiming at harmonizing with his movements. I did not experience any explicit scaffolding guidance. Rather I had to follow and take in the sensation that was enacted in the interaction and find a way to make them mine. He was the experienced other, practising as normal. In this setting, I was the newcomer who had incorporated 'just enough' capabilities to be able to form part of the interaction offered. He was showing me an honour and I found myself in a position striving to live up to this honour while being aware how this unique interaction cultivated my sensations while exercising the techniques.

This was my second round of practice that morning. I can see from my notes that the tiny muscles deep inside my hip joints 'screamed' and halfway through the practice I was not sure I could complete the session. I had to let go of the tension (to 'survive') while staying focused at the same time. The more I let go of insisting on directions and accuracy, the easier I felt it was to move his centre. At the time, I strongly felt I had to rely on my peripheral vision in the interaction to be able to blend with the flow and the energy of his movement. His expression was (as mine, I presume) stone-faced. No expression until the end of the practice. When I thanked him by remembering to quickly move to *seiza* and bow a bit lower than usual to show my gratitude and respect, his face opened into a big and friendly smile. He never said a word during the practice. There had just been a nod or a minimal headshake here and there. Taka-sensei shared with me the tacit insight of his way of practising aikido by guiding my movements. Throughout the practice with Taka-sensei, I aimed at harmonizing with his energy and the way he moved – at the same time he seemed to be both with my movements and ahead of them. Taka-sensei thereby offered me a route to discover new aspects and depths of the techniques we practised as well as of the way movements connect within my body.

Skilled practices of interacting in aikido

In *Phenomenology of Perception*, Merleau-Ponty has highlighted how our lived body is our potentiality to a certain world (1962, 106, 148–9, 153). We are anchored in our body as a moving and pre-reflective seat of power, an 'I can' (1962, 153). In his elucidation of this potentiality, Merleau-Ponty stresses that the other person presents a familiar way of dealing with the world. The body of the other person presents a miraculous prolongation of my intention already on the level of operative intentionality. I sense the intentions of others immediately in their actions, and act in coordinated ways along with the movements initiated. Similarly, others act in accordance with the pre-reflective operative intentionality I enliven. When practising aikido, I participate in a particular world of social interactions which are centred around couple-based skills but reach beyond the two practising together. The two practising together are part of a particular world which already includes a 'being-with' (Gallagher 2008; Krueger 2011). Being in the dojo, connecting to the aikido heritage and grand stories while enacting certain codes of conduct, surrounds and includes the subjects and their actions. According to De Jaegher and Di Paolo (2007) and Fuchs and De Jaegher (2009), sense-making is effectively a process of the coordination of an agent's values and intentions with its environment (see also McGann 2014, 5). There is a lot more than the here and now of the interaction contributing to and forming part of the sense-making enacted. Interacting in aikido includes finding partners to practice with, knowing codes of conduct enabling and constraining the way practice can be enacted and attuning carefully to the other and to the sensations of one's own body. It is a complex affair where sense-making reaches beyond the here and extends before the now of the two practising together.

Cultivating one's aikido skills through the experienced others

Phenomenological and enactive descriptions have specifically shown how fine-tuned adjustments characterizing micro-levels of physical interaction are coregulated between two or more bodies (De Jaegher and Di Paolo, 2007; Fuchs and De Jaegher 2009). Continuing Merleau-Ponty's work, they indicate that the meaningful connection between perception and movement cannot be reduced to an affair of perception–action cycles of singularized bodies. Rather, the perception in which we are already acting forms part of complexities of interlinked perception–action cycles. However, as McGann (2014) makes aware, these descriptions seem to focus on the experiential perspective of the agent and one-to-one interactions – like, for example, when we coordinate our actions to pass another person in a corridor. Accordingly, descriptions and the examples given by Fuchs and De Jaegher (2009) do not easily translate to the complex practices which demand us to interact with several others and on different levels at the same time.

Despite this critique (e.g. Baggs and Chemero 2021; McGann 2014) it is important to emphasize that Fuchs and De Jaegher (2009) bring valuable attention to how intentionality is shaped and enacted in interaction processes. They specify that the process of interaction between two agents is driven forward in a continuous fluctuation between synchronized, desynchronized and in-between states of intentionalities. The intentionality of the interactors might decentre so that, in synchronized moments, the agents involved will experience the interaction process gaining a 'life of its own' (2009, 471). This can, for example, be the case when dancing together. As I have argued elsewhere, the couple who improvises the Argentinean tango can experience that at times, when the dancing works especially well, the dance is experienced as dancing them. In these moments, the intentionality of the next step takes shape in the very life of their interaction (Ravn 2019). The tango dancers value and strive towards achieving such synchronized interaction for greater parts of their interaction. In a comparable way, the ideal of harmonizing in the dyadic skill of two people practising together in aikido requires them to strive to synchronize. However, the skilled practitioner engaged in the antagonistic kind of interaction in aikido cannot strive (only) for synchronization. Synchronization is not the end goal but works as a foundation for taking control (as *tori*) and for tapping in and learning from the experienced others. As exemplified in my training with Taka-sensei, *tori* will strive to harmonize to be slightly ahead of the intentionality of *uke's* movement, and *uke* will strive to both be with and give in to the interaction so that the experienced other can take charge and be ahead of *uke*'s motor intentionality throughout. The complex part of the crocodile story, however, is that Taka-sensei was ahead of my movement both as *tori* and *uke*. As *uke*, he followed me, but only to the degree he intended while finding a way to cultivate my technique. I had to know, through my own ongoing attunement to the practice, how to move to harmonize with Taka-sensei in a practice where he both harmonized and was ahead of the operative intentionality of my movement.

Conclusion

Based on ethnographic data, I have shown that the antagonistic practice of aikido is far more complex than the interactional dynamic between the two practising together in the here and now. The enaction and transmission of bodily knowledge, unfolding in aikido-specific sense-making practices, are interwoven processes in which practitioners coordinate values and intentions with the ecology of aikido. Incorporated techniques, codes of conducts, the sensation of the *gi*, belt, *hakama* and the bare feet on the aikido mats can interweave with features of the practitioner's bodily style when forming part of the aikido-specific interaction. In my analysis, I have specified how the aikido practitioner cultivates their skills *through* their partner's movements. Thus, it is through the interactive connection with Sensei's experienced body that I have learned what it means, in a bodily sense, to be centred in a relaxed way, to move from my hip, to find the right timing and so on, while developing and cultivating my aikido techniques. Practising with Taka-sensei – harmonizing with a highly experienced other – I exercised my skills and competences on the premises of harmonizing. Taka-sensei both seemed to be harmonized and in sync with the coordinated interaction – yet also ahead of my intentionality. Accordingly, I suggest that cultivating one's skills in a specialized practice such as aikido demands an intentional engagement operating on multiple levels at the same time: harmonizing (aiming at being in sync) while striving to be ahead and in control or harmonizing (aiming at being in sync) while striving to be ahead and in control or harmonizing one's attacking movement by listening and giving in to the interaction, decentering intentionality to be led.[5]

Notes

1 Throughout the article I use the notion 'attacking'. However, it should be noted that in aikido practices *uke* is first of all understood as a partner helping *tori* to practice their technique. It would be a mistake to think of the aikido attack as if, for example, it was aimed at obscuring *tori*'s technique by using unorthodox attacks.
2 In the following, when using Sensei, with a capital 'S', it refers to sensei Victor Merea.
3 To see the technique demonstrated, visit https://www.youtube.com/watch?v=uwktFvFGirk.
4 'Taka'-sensei is a fictional moniker to anonymize the practitioner.
5 I would like to thank Jesús Ilundain, Kath Bicknell, John Sutton and an anonymous reviewer for their fruitful comments and valuable suggestions throughout the process of improving this contribution. I would also like to thank aikido-aikikai practitioners who have generously shared their embodied insights in and through practising. Not least, thanks to Sensei Victor Merea for two decades of patience with me as his aikido student.

References

Baggs, E. and A. Chemero (2021), 'Radical Embodiment in Two Directions', *Synthese 198* (9), 2175–90.
De Jaegher, H. and E. Di Paolo (2007), 'Participatory Sense-Making – An Enactive Approach to Social Cognition', *Phenomenology and the Cognitive Sciences* 6: 485–507.
Downey, G. (2008), 'Scaffolding Imitation in Capoeira: Physical Education and Enculturation in an Afro-Brazilian Art', *American Anthropologist* 110 (2): 204–13.
Fuchs, T. and H. De Jaegher (2009), 'Enactive Intersubjectivity: Participatory Sense-Making and Mutual Incorporation', *Phenomenology and the Cognitive Sciences* 8: 465–86.
Hammersley, M. and P. Atkinson (2007), *Ethnography: Principles in Practice*. 3rd ed. London; New York: Routledge.
Gallagher, S. (2008), 'Philosophical Antecedents of Situated Cognition', In P. Robbins and M. Aydede (eds), *Cambridge Handbook of Situated Cognition*, 35–52. Cambridge: Cambridge University Press.
Kohn, T. (2003), 'The Aikido Body: Expressions of Group Identities and Self-discovery in Martial Arts Training', In Dyck, N. and Archetti, E. P. (eds), *Sport, Dance and Embodied Identities*, 139–56. Oxford: Berg.

Krueger, J. (2011), 'Extended Cognition and the Space of Social Interaction', *Consciousness and Cognition 20* (3): 643–567.

McGann, M. (2014), 'Enacting a Social Ecology: Radically Embodied Intersubjectivity', *Frontiers in Psychology*, 5: 1321.

McGann, M. (2020), 'Convergently Emergent: Ecological and Enactive Approaches to the Texture of Agency', *Frontiers in Psychology*, 11: 1982.

Merleau-Ponty, M. (1962), *The Phenomenology of Perception*. London and New York: Routledge, (reprinted 1998).

Mingon, M. and J. Sutton (2021), 'Why Robots Can't Haka: Skilled Performance and Embodied Knowledge in the Maori Haka', *Synthese*. https://doi.org/10.1007/s11229-020-02981-w.

Palmer, W. (2002), *The Practice of Freedom – Aikido Principles as a Spiritual Guide*. Berkeley: Rodmell Press.

Ravn, S. (2019), 'Improvisation and Argentinean tango – playing with Body Memory', In Midgelow, V. (Ed.) *The Oxford Handbook of Improvisation in Dance*, 297–310. New York: Oxford University Press.

Ravn, S. and H. P. Hansen, H (2013), 'How to Explore Dancers' Sense Experiences? A Study of How Multi-sited Fieldwork and Phenomenology Can be Combined', *Qualitative Research in Sport, Exercise and Health 5* (2): 196–213.

Sutton, J., D. J. F. McIlwain, W. Christensen and A. Greeves (2011), 'Applying Intelligence to the Reflexes: Embodied Skills and Habits between Dreyfus and Descartes', *Journal of the British Society for Phenomenology* 42(1): 78–103.

Westbrook, A. and O. Ratti (1973/2001), *Aikido and the Dynamic Sphere: An Illustrated Introduction*. Boston: Tuttle Publishing.

Ueshiba, M. (2005), *Progressive Aikido: The Essential Elements*. New York: Kodansha America, Inc.

7

Musical agency and collaboration in the digital age

Tom Roberts and Joel Krueger

Introduction

In 2019, the musician Holly Herndon released her third full-length album, *Proto*. In addition to input from two other human artists, the album had a fourth collaborator: an artificial neural network named *Spawn*. The software had been trained over several years to generate and manipulate the cavernous choral soundscapes that brought *Proto* widespread critical acclaim. *Spawn's* role in each stage of the music-making process was neither completely predictable nor completely under Herndon's control; her vocal contribution – its tone, pitch, rhythm and dynamics – was often novel, original and surprising.[1] Herndon describes *Spawn* as 'a performer … an ensemble member. So I would say that I collaborated with a human and an inhuman ensemble' (Funai 2019).

Here, we consider how seriously we ought to take assertions like this one. Can we *really* conceive of AI systems as legitimate collaborators in the skilled project of making art? Do they have the kinds of creative agency, autonomy and expressive power that characterize membership of an artistic ensemble?[2]

In the next section, we rehearse some reasons why there has been a reluctance to give affirmative answers to these questions – why, that is, computational systems have been taken to have an impoverished status, lacking capacities essential to true artistic agency (see Boden 2007). Following this, we explore the view that even when attributions of creativity and autonomy to artificial systems are not *literally* true, they can instead be *fictionally* true. Those who work alongside generative systems like *Spawn* and those who enjoy the musical fruits of such collaboration are participants in an elaborate game of make-believe, wherein the non-human contributor is imaginatively conceived as being a real improviser, a real singer, a real musician. Taking this line allows us to give credence to testimony like Herndon's, and to better understand the production and appreciation of music that has a partially non-human origin.

Musical agency

Why might *Spawn* be regarded as deficient, relative to the skills and capacities of more traditional makers of music? Here, we sketch three related characteristics that lie at the heart of musical agency in familiar contexts: embodiment, emotional expression and autonomy. Human agents typically exemplify these dimensions in the course of making music, to a greater or lesser degree, but it is hard to see how they might be manifested by a robot, an algorithm or a neural net.

Firstly, the performance of traditional acoustic music is an *embodied*, *energetic* and *visceral* affair. Instruments are blown, struck, plucked, strummed and twanged with a rhythm and vitality that reflects physical engagement with the music. The musician and her instrument are in motion together; motion shaped by grip, posture, muscle and breath. When an ensemble of musicians plays in unison, moreover, their bodies attune to one another in the service of a collective aesthetic aim (Clayton et al. 2020). And of course the body itself may be an instrument – the voice, the stamping of the feet, the clapping of the hands.

Secondly, music can be a vehicle for *emotional expression*: a powerful tool for articulating the affective states of a listener, composer or performer. Facility with a musical instrument can expand and enhance an agent's expressive repertoire, giving her a new language with which to convey her feelings. Pitch, loudness, rhythm and timbre can enrich the musician's emotional vocabulary. And music's affective content is carried to the ear of the listener, too, who may in turn be moved, saddened, uplifted or called to action.

Thirdly, music typically arises from acts of *creative autonomy*, governed by the artist's choices and intentions. Although not every note or phrase is the product of conscious deliberation, the artist controls the overall process of conception, composition and performance and bears responsibility for the music's final form.[3] Various layers of intentional input are possible: a composer may devise and transcribe the melody, for example, and arrange parts for the orchestra to follow. An individual player can choose the tempo and dynamics of a piece, and when to diverge from or embellish a score. Sometimes these choices are made collectively, in advance or on the fly, in discussion or rehearsal.

Our claim is not that every musical performer, nor every member of an ensemble, must always exhibit each of these three features to a high degree. Sometimes, after all, a musician may simply follow a score and submit her own agency to that of the conductor or band leader; and sometimes a performance may be a tightly controlled technical feat, with little room for emotional colour.

What we do suggest is that embodiment, expressivity and autonomy are characteristic aspects both of how we conceive of musical agency and of how we experience musical performance in a range of ordinary cases. We *hear* music, that is, as the product of an act of singing or playing, shaped by the artist's intentional agency and delivered through embodied, expressive behaviour. This agential character, moreover, shows up in our appraisal of a work or performance as an *achievement* (Huddleston 2012; Roberts 2018). For instance, a technical achievement of dexterity, breath control or coordination, or as the virtuous product of originality, honesty or insight. Music is not only an unfolding pattern of sound, it is the result of effort, intention and expertise; and it is conceived of, perceived and evaluated in these terms by the audience.

In sum, musical agency has several facets that come in degree: it involves a package of features that implicate intention, expression and cognition, where these features are most typically borne by embodied human subjects.

What do these remarks tell us about *Spawn* and her kin? They may encourage a pessimistic position regarding the creative and expressive powers of artificial systems. A neural network housed in a box is not alive, inhabits no organic body, lacks projects and concerns, feels no emotions and has no evaluative capacities. It cannot tap its fingers; sway to a beat; or feed on the energy of its bandmates. While it has generative capabilities, it cannot select or refine its own outputs on the basis of their aesthetic interest,[4] let alone commit to artistic projects that have wider cultural or political resonance. In the absence of conscious inner states, it is unable to express a concernful perspective – to communicate feelings of loss, say, or to voice its joyful triumph. In accepting this pessimism, one might concede that talk of AI artistry, collaboration, expression and creativity is simply false and misleading, and to reserve these terms for full human agents. Yet this concession appears to be at odds with what we hear from artists like Holly Herndon, who seem willing to attribute a degree of artistry, agency and autonomy to their artificial collaborators.

Fictional agency, fictional artistry

We propose that an alternative, *fictionalist* approach can illuminate the artistic and appreciative practises that grow up around AI-driven music. The fictionalist view enables us to say that it can be advantageous for an artist or listener to engage in the fictional pretence that there is AI musical agency – ranging from performance and interpretation to full creative composition – even if we accept that this is not literally true.

In the philosophy of mind, fictionalism is the view that even when we attribute inner mental states to other humans, we are engaging in a complex act of pretence. We don't sincerely judge that there are internal beliefs and desires, for example, but it is extremely productive, for the purposes of explanation and prediction, to treat one another as though we have them. The imaginative game of make-believe in which we collectively participate is a false but highly useful tool for navigating the interpersonal world. Fictional make-believe is a more involved mental process than the 'detached imagining' we perform when, for example, we conceive an abstract philosophical thought experiment. Make-believe has deeper ties to behaviour: we interact physically with the material 'props' of the fictional setup (Walton 1990) and these interactions in turn generate new imaginings, governed by the rules of the game we are playing (Toon 2016).

While we have no wish to defend fictionalism's systematic anti-realism here,[5] we will argue that there is value in applying it to the particular domain of artificial systems, including those that appear to be operating creatively. The claim is that **even if**, strictly speaking, artificial musical collaborators lack an autonomous, expressive point of view, it can be fruitful for an artist to participate in the fiction that they *do* exhibit this richer agential status.

Notice, as a preliminary, how natural it is to describe in fictionalist terms our anthropomorphization of entities such as robots and videogame characters. When we are presented with an on-screen humanoid or a mechanical creature whose behaviour appears goal-directed, intelligent or otherwise non-random, we are strongly inclined to react to them as though they were a psychological agent. We imaginatively entertain that the cute robot dog is a friendly pet who wants to play; and that the zombies in the videogame have a murderous intent and a thirst for revenge. If pushed, we would surely deny that ascriptions like these are literally true; but it is nonetheless *part of the fun* that – for a while at least – we act as though they are.

Within a fiction, we might consider even a rather rudimentary artificial system to have a quite sophisticated mental life – short and long-term plans, memories, preferences, moods and so forth.[6] And our own psychological and behavioural reactions are shaped by the role we adopt in the fiction, too – in our displays of sympathy, attachment or solidarity towards the robot pet we have been raising, for example, or the fear and hostility we feel towards the antagonists in the videogame. Entering into the make-believe with enthusiasm and goodwill, we suggest, is often the best way to make the most of the interactive opportunities afforded by novel technologies. Treating a virtual or artificial entity *as if* it had folk-psychological states makes it possible to form certain new relationships with that entity – to consider it a friend or foe, for instance – and allows us to predict, explain and interpret the entity's behaviour without attending to its underlying physical or computational basis,[7] in real time, much like we do when we interact with human agents.

The phenomenology of fictional agency

Using Herndon as a case study, we now consider some ways of conceiving how we might collaborate with AI systems.[8] Fictionalism can help to make sense of some illuminating tensions in how Herndon seems to experience and describe her collaboration with *Spawn*. It can also help to better understand why *Spawn* – and AI systems more generally – offers collaborative relationships that are richer, and potentially more artistically productive, than those afforded by other non-human resources sometimes brought into the music-making process.

Performative and compositional collaboration

It is not uncommon to speak of non-human resources as involved in the creative music-making process (de Mori 2017). Indigenous peoples may describe songs as originating from guardian or ancestral spirits; Western composers such as Brahms portray themselves as a conduit for music that flows directly from God; musicians like Brian Eno use card-based methods ('Oblique Strategies') to prompt creative thinking; while Pauline Oliveros's 'Deep Listening Band' performed in subterranean vaults that shaped their music's reverberating character.

In these cases, non-human resources are said to play an important role in animating the creative process. But it is unlikely that the resources would be described using the agential vocabulary of thinking, feeling or intending.[9] However, collaborations with artificial systems like *Spawn* appear to more readily invite folk psychological attributions. One reason for this is that they furnish practical, experiential and temporal (both synchronic and diachronic) forms of collaboration that are richer and more complex than those offered by other non-human resources – forms of collaboration, that is, that *feel* closer to engaging with a human agent than does 'collaborating' with, say, a deck of cards. Whereas the latter may provide a useful stimulus – a cryptic remark on an Oblique Strategies card may prompt an insight that helps to overcome a creative impasse – their causal input remains limited, and their interactive possibilities static. In contrast, AI systems like *Spawn* offer deeper forms of engagement closer to the collaborative dynamics that unfold between human agents. Not only are they structurally more complex – in being *iterative*, *reciprocal* and *temporally extended* – they can appear to bear the hallmarks of agency and mindedness, seeming to have a musical voice of their own.

Herndon and human colleagues first train *Spawn* by creating data sets made up of Herndon's own voice and those of an ensemble. They then feed sonic building blocks (vocals, percussive elements, etc.) into *Spawn*, who draws on these data to sing over these building blocks – often in unpredictable and surprising ways. Herndon then splices this *Spawn*-produced output into tracks (sometimes recording more vocals in response), or feeds her manipulations back into *Spawn* in order to generate further outputs.

Despite the important role *Spawn* plays in the creative process, Herndon is clear that she is not sentient: 'I don't see Spawn as a human baby. I see Spawn as an artificial intelligence baby .... It's something that can surprise and can have the feeling of creativity and ingenuity, but there's no consciousness yet' (Friedlander 2019).

Words like these suggest that our fictionalist characterization is a complicated matter. What, then, is the value in applying fictionalism to creative- and artistic-looking procedures like music-making? One significant value is *phenomenological*. Fictionalism can illuminate the felt character of the sorts of collaborative experiences artists like Herndon describe, including what may initially appear to be some puzzling features of her descriptions of these experiences. For, despite her protests to the contrary, some things Herndon says suggest that she does, in fact, adopt an anthropomorphic fictionalist stance towards *Spawn* when collaborating with her. This is why their collaborations work as effectively as they do.

Recall first that Herndon is comfortable speaking about her partnership with *Spawn* as a genuine collaboration. She says:

> I consider Spawn as a performer, as an ensemble member ... I certainly consider those collaborations. When you write a score, then somebody reads it, human or inhuman, there's an interpretation happening there. Things always come out slightly different than when you imagined it. That's how I've used Spawn as a performer. It's collaborative in that sense.
>
> (Funai 2019)

Here, Herndon characterizes her collaborative relationship with *Spawn* in terms of *interpretation* and *performance*. *Spawn* performs her take on the music that Herndon creates. Herndon affirms this characterization elsewhere, saying that 'we see *Spawn* as an ensemble member, rather than a composer. Even if she's improvising, as performers do, she's not writing the piece. I want to write the music!' (Hawthorne 2019).

Herndon is clear that although she herself is the composer responsible for the choices and intentions behind the music,[10] *Spawn* may be said to interpret the piece by manipulating some of its elements – the way a musician may manipulate, say, the tempo or dynamics of a score when performing live in response to the audience or their own aesthetic impulses. Talk of 'reading' and 'interpreting' the score, we suggest, may already be non-literal – terms like these suggest a cognitive sophistication that we might be unwilling to attribute to *Spawn*. If so, the fictionalist view permits us to say that Herndon is not mistaken or speaking falsely when she uses such language; instead, it is a make-believe that *Spawn* is an ensemble member.

Sometimes, moreover, Herndon also seems comfortable describing her collaboration with *Spawn* not just in terms of *performance* but also of *composition*. In other words, *Spawn*'s role is felt to be more than just an expressive vehicle articulating Herndon's pre-formed vision; rather, she (*Spawn*) contributes something more substantial, much closer to creative agency. Herndon acknowledges that in the case of *Spawn*, the boundary between performance and composition can blur:

> There's often this extreme hierarchy between composer and performer ... I'm not saying this is non-hierarchical – my name's on it, I'm choosing which performances land on the record – but ideas aren't generated in a vacuum. The idea of one person being the entirety of something is just really limited.
> (Hawthorne 2019)

What Herndon seems to suggest here is that the creative agency driving the music-making process is not limited to one causal origin. It is instead a collective enterprise, something distributed across multiple agents – one of whom happens to be non-human.

Similarly, Herndon tells us elsewhere:

There is some improvisation that happens when Spawn interprets something that I write. It's not a binary between composing and performing. There is an entire gray area of interpretation and the improvisation. However, I prefer to stay on the end of maintaining the composition ... I like to maintain that autonomy and that agency of being able to grow and change my aesthetic and change my form ....

(Funai 2019)

This quote captures the tension in how Herndon seems to experientially relate to *Spawn*. On one hand, Herndon is keen to maintain a grip on creative agency and authorship, conceiving of *Spawn's* role in terms of (mere) performance. However, on the other hand, she also seems to concede that *Spawn* generates goods that are somehow essential for driving the creative process: resources that contribute to her own growth as an artist.

This idea of modulating her 'aesthetic and form' in order to animate the creative process is found in yet another description: 'I'm singing through a system I've made [i.e. *Spawn*]. I can morph between human and animal and digital. I can sing through plants' (Hawthorne 2019).[11] The 'morphing' Herndon describes is, as we'll see below, a *modulation of agency* – a transformation, guided and scaffolded by *Spawn's* ongoing input, that helps her get into the creative space needed to compose her distinctive music.

How, then, should we understand this tension in Herndon's reports? Is her collaborative relationship with *Spawn* primarily performative, compositional or somehow both? Here, fictionalism gains further traction. We propose – as the best interpretation of available evidence on Herndon's attitudes and practice – that when making music, Herndon adopts a fictionalist stance towards *Spawn*. Although she clearly knows that *Spawn* is not a conscious subject, she nevertheless treats *Spawn as if* she has a mental life – as if she is a kind of agent with aesthetic beliefs, desires, intentions, etc. – in order to temporarily become part of a larger structure of collaborative agency.[12] By adopting a fictionalist stance, Herndon allows *Spawn* to take over aspects of performance and composition to contribute novel (and often unexpected) goods that open up previously unseen creative pathways. Incorporating *Spawn* into the creative process in this way allows Herndon to experiment with temporary agencies ('I can morph between human and animal and digital'); this experimentation is a central part of the music-making process. We now consider the phenomenology of this experimentation in more detail.

Experimenting with (fictional) agencies in music

Note first that using music to experiment with our agency is not something confined to music-making with artificial systems. We regularly do something like this when listening to music, too. There is a sense in which we *enter into* music (Krueger 2009). We experientially inhabit it and let it take over and govern different aspects of our agency. Briefly considering how so will shed light on our experimentation with fictional agencies when collaborating musically with artificial systems.

There is a tight link between the form of our musical engagements and the way we experience and manipulate different aspects of our agency within these engagements (Krueger 2014). For example, several scholars defend the idea that musical experience can involve the presence of an imagined 'other' (Levinson 2006), a 'persona' (Cochrane 2010) or 'virtual agent' (Leman 2007) with whom we identify when immersing ourselves in a musical work.

Music is also a powerful resource for the construction of the self and social relationships. As DeNora (1999) puts it, music is a 'technology of the self' – a resource or 'material that actors use to elaborate, to fill out and fill in, to themselves and to others, modes of aesthetic agency and, with it, subjective stances and identities' (p. 54). Varieties of musical practices central to everyday life are, in this way, tied to the construction, experience and manipulation of our agency.

For our purposes, the key point is this: music furnishes resources that allow us to experiment, in various ways and at multiple timescales, with *forms of agency* – including, in the context of AI-driven music-making, *fictional* agencies. When composing, we suggest that Herndon treats *Spawn as if* she has a mental life (beliefs, desires, intentions, creative impulses, etc.), in order to bring her more deeply into the creative process, to *feel* like *Spawn* is more deeply involved – and in so doing, generate new interpretive and compositional possibilities. However, this fictionalist stance also shapes Herndon's *self-experience*. By allowing herself to become drawn up into this larger collaborative structure – by offloading part of the creative process onto *Spawn* – Herndon can, in turn, experiment with *her own* agency. She can 'morph between human and animal and digital' and 'sing through plants' as she temporarily inhabits new creative spaces opened up by this organic-digital collaboration.

To develop this idea further, we can use Nguyen's (2019) work on games and agency. Nguyen argues that a similar process unfolds when we play games, especially computer games offering visually immersive worlds and rich story- and character-driven narratives. For Nguyen, games specify modes of agency for players to adopt: their rules, practices, goals and supporting abilities 'shape the agential skeleton which the player will inhabit during the game' (2019, 423). For example, undertaking projects, tasks or quests alone or with others; developing a character's skills, abilities or motivations; interacting with non-player characters to advance the storyline, etc., allows players to take on alternative agencies in a controlled and limited way. Players fictionally become things they're not and do things they can't normally do because the game space furnishes resources supporting this sort of agential transformation. They can engage in these transformative practices, Nguyen argues further, because human agency is not fixed. It turns out to be 'modular and moderately fluid. We have the capacity to set up temporary agencies, layered within our larger agency, and submerge ourselves within them' (2019, 426).

There are, of course, structural and phenomenological differences between playing games and making music with AI. One difference concerns the respective *aims* of these activities. When playing games, a central part of the enjoyment experience is the experience of *striving*. Game designers not only create the world in which individuals will act but also structure their practical agency – their abilities, goals and values (2019, 438). Enjoyment of games is tied to a *balanced* striving experience: too much freedom and the game will become tedious; too little and it is frustrating. We take on temporary agencies for the sake of the intrinsic value of the experience of struggling within the gameworld; we enjoy the strenuousness of the play, the tension, uncertainty and (assuming we achieve our goal) release of finally realizing the fruits of our striving.

In the case of collaborating musically with artificial systems, it's unclear that striving plays the same role. *Spawn* is set up to contribute novel and unpredictable responses, and to challenge her human collaborators, by forcing them to respond to her outputs in unanticipated ways. This tension and uncertainty fuels the creative energy driving the music-making process. It may be thought of as a kind of striving – in much the way that improvising with a new musical partner (or partners) involves a kind of striving as individuals work to get into a groove with one another by learning to adapt and respond to each others' idiosyncratic styles. However, it's unlikely that Herndon et al. are interested in the intrinsic value of this striving itself. Rather, the striving is a means towards some further end – namely, to make music. In this context, the striving experience has an instrumental value that distinguishes it from playing games.

Conclusion

Fictionalism helps to illuminate why Herndon is motivated to adopt a make-believe stance towards *Spawn*. By treating *Spawn as if* she is an agent with (some degree of) creative autonomy, as if her input is to be taken as seriously as that of a human collaborator, Herndon generates the aesthetic tension, the *striving*, needed to drive the music-making process. Crucially, this striving has a rich diachronic and reciprocal character that distinguishes it from the stimulus–response structure that characterizes other non-human contributions to music, such as Eno's Oblique Strategies cards or Oliveros's underground caverns. *Spawn* provides ongoing (and often unpredictable) resources that give Herndon a felt sense that *Spawn* is a participatory member of the creative process, generating ideas and aesthetic energy that drives the process along. Part of this feeling also seems to arise from the *interactive* possibilities *Spawn* affords. Herndon can play with and manipulate *Spawn's* output – riff on it – and potentially feed her riffing back into *Spawn* in order to generate *further* output. This interactive and *iterative* dynamic helps understand why Herndon may be inclined to adopt a fictionalist stance towards *Spawn*, despite her firm insistence that *Spawn* is not sentient.

Fictionalism can also help illuminate another dimension of the collaborative process. By adopting a fictionalist stance, Herndon is also able to temporarily inhabit agential structures – Nguyen's 'agential skeleton' – that enable her to experiment with and explore structures of *her own* agency. By singing through the technological resources *Spawn* provides, she can generate and inhabit richly textured soundworlds that would otherwise be unachievable. She can experiment with different modes of creative agency and gain insight into her own creative process as she responds to what *Spawn* feeds back to her.

For Herndon, then, this complicated collaborative relationship with *Spawn*, and the forms of agential experimentation it affords, is not something that alienates her from her humanity. Rather, it *affirms* it. She tells us that 'technology should allow us to be more human together rather than alienating us further. So many of the products and so many of the habits that we have with our technology pushed us towards alienation. But really, it could free us up to be more human and more emotional together by taking some of the work, essentially' (Funai 2019). In this context, fictionalism is one source of such freedom.[13]

Notes

1 Herndon and her team use female pronouns for Spawn.
2 While our interest in these questions is philosophical, it is not difficult to see that they may have legal or economic implications – can an AI claim intellectual property rights?
3 This is true even with techniques of composition that introduce elements of randomness or chance, such as Iannis Xenakis's stochastic or aleatoric music, or the generative music created by electronic musicians like Autechre, Keith Fullerton Whitman and Emily A. Sprague.
4 For discussion of the selection phase of creativity, see e.g. Boden and Edmonds (2009), Wheeler (2018).
5 One of us has argued against a core assumption motivating fictionalist approaches to other minds: that folk psychological pretence is necessary because we have no way of directly accessing others' mental states (e.g. Krueger 2012, 2018). We also note that, as a philosophical hypothesis about our attributions of mental states, fictionalism is constrained by and consistent with scientific evidence of the kinds mentioned in the text. Debates about the basis of such attributions arise, like many in the philosophy of mind, because the evidence from psychology and neuroscience does not fully settle the issue.

6 There is some evidence that we are more inclined to attribute cognitive than emotional states to AI (e.g. Bakpayev et al. 2020).
7 That is, by adopting the intentional rather than the physical stance (Dennett 1987).
8 Herndon is not the only electronic musician who collaborates with an artificial system. We choose to address her work here due to the wide-ranging and nuanced descriptions she has given of her collaborative creative process. We acknowledge the limitations of relying on the testimony of a single artist, offered in a non-academic context, and in future work we hope to engage with further musicians and performers.
9 The case of God, or guardian or ancestral spirits, is more complicated since individuals are inclined to speak of these entities in folk psychological terms. Since our focus is on material artefacts and technologies of music-making, we do not consider these cases further.
10 Moreover, we might interpret her insistence that 'I want to write the music!' as implying that to attribute creative authorship to *Spawn* would somehow be inauthentic (Boden 2007).
11 By 'singing through plants', Herndon means that *Spawn* allows her to experientially inhabit and manipulate field recordings in real time, using her voice.
12 Something like this is what Herndon seems to have had in mind when, during a recent online video discussion along with her collaborator, Mat Dryhurst, she described this experience as 'getting lost in the romance' of making music with *Spawn*. Holly Herndon and Mat Dryhurst in conversation. Everywhere it is Machines series, School of Performing & Digital Arts, Royal Holloway, University of London, 1 April 2021.
13 Many people provided helpful feedback on previous drafts of this chapter. We are particularly grateful to Lucy Osler, Giovanna Colombetti, Adrian Currie, Juan Diego Bogotá, John Sutton, Kath Bicknell, Emily Cross, Kohinoor Darda, Michael Wheeler, Ian Maxwell and participants in the 'Collaborative Embodied Performance Work-in-Progress' workshops and University of Exeter 'Culture and Cognition' reading group.

References

Bakpayev, M., T. H. Baek, P. Van Esch and S. Yoon (2020), 'Programmatic Creative: AI Can Think but It Cannot Feel', *Australasian Marketing Journal (AMJ)*, April. https://doi.org/10.1016/j.ausmj.2020.04.002.
Boden, M. A. (2007), 'Authenticity and Computer Art', *Digital Creativity* 18 (1): 3–10.

Boden, M. A. and E. A. Edmonds (2009), 'What Is Generative Art?' *Digital Creativity* 20 (1–2): 21–46.

Clayton, M., K. Jakubowski, T. Eerola, P. E. Keller, A. Camurri, G. Volpe and P. Alborno (2020), 'Interpersonal Entrainment in Music Performance', *Music Perception* 38 (2): 136–94.

Cochrane, T. (2010), 'Using the Persona to Express Complex Emotions in Music', *Music Analysis* 29 (1–3): 264–75.

De Mori, B. B. (2017), 'Music and Non-Human Agency', In *Ethnomusicology: A Contemporary Reader*, Volume II, ed. Jennifer C. Post, 181–94. London: Routledge.

Dennett. (1987), *The Intentional Stance*. Cambridge, MA: MIT Press.

DeNora, T. (1999), 'Music as a Technology of the Self', *Poetics* 27 (1): 31–56.

Friedlander, E. (2019), 'How Holly Herndon and Her AI Baby Spawned a New Kind of Folk Music', *Fader*, 21 May. Available online: https://www.thefader.com/2019/05/21/holly-herndon-proto-ai-spawn-interview.

Funai, M. (2019), 'Holly Herndon on Merging the Worlds of Music and AI', *Dropbox Blog*, October 10. Available online: https://blog.dropbox.com/topics/our-community/holly-herndon-interview (accessed 21 June 2021).

Hawthorne, K. (2019), 'Holly Herndon: The Musician Who Birthed an AI Baby', *The Guardian*, 2 May. Available at: https://www.theguardian.com/music/2019/may/02/holly-herndon-on-her-musical-baby-spawn-i-wanted-to-find-a-new-sound (accessed 21 June 2021).

Huddleston, A. (2012), 'In Defense of Artistic Value', *The Philosophical Quarterly* 62 (249): 705–14.

Krueger, J. (2009), 'Enacting Musical Experience', *Journal of Consciousness Studies* 16 (2–3): 98–123.

Krueger, J. (2012), 'Seeing Mind in Action', *Phenomenology and the Cognitive Sciences* 11 (2): 149–73.

Krueger, J. (2014), 'Affordances and the Musically Extended Mind', *Frontiers in Psychology* 4: 1003.

Krueger, J. (2018), 'Direct Social Perception', In Albert Newen, Leon de Bruin and Shaun Gallagher (eds), *Oxford Handbook of 4E Cognition*, Oxford: Oxford University Press.

Leman, M. (2007), *Embodied Music Cognition and Mediation Technology*. Cambridge, MA: MIT Press.

Levinson, J. (2006), *Contemplating Music*. Oxford: Oxford University Press.

Nguyen, C. T. (2019), 'Games and the Art of Agency', *The Philosophical Review* 128 (4): 423–62.

Roberts, T. (2018), 'Aesthetic Virtues: Traits and Faculties', *Philosophical Studies* 175 (2): 429–47.

Toon, A. (2016), 'Fictionalism and the Folk', *The Monist* 99 (3): 280–95.
Walton, K. L. (1990). *Mimesis as Make-Believe: On the Foundations of the Representational Arts*. Cambridge, MA: Harvard University Press.
Wheeler, M. (2018), 'Talking about More than Heads: The Embodied, Embedded and Extended Creative Mind', In B. Gaut and M. Kieran (eds), *Creativity and Philosophy*, 230–50. New York: Routledge.

Commentary: Embodied learning within embodied communities

Emily S. Cross

The four chapters in this section take us from sturdy gym mats where our authors learned to perfect strong and stable handstands or honed the aikido ideal of harmonizing with an opponent, to deep below the water's surface, where breath-holds became successively, achingly longer, and finally into the realm of the seemingly surreal, where artificial neural networks became musical collaborators. The authors shared authoritative and reflective accounts of embodied learning, collaboration and socially scaffolded cognition. Greg Downey's suggestion that 'the path to expertise is signposted by the community' struck me as particularly astute, and even poignant given the challenges to sharing and collaborating with one another in an embodied context that have characterized human life for almost everyone on earth since the arrival of COVID-19 in early 2020 (just over a year before the time of writing). In this commentary, I reflect upon some lessons, observations and insights in light of the pandemic-induced (and hopefully temporary) shift away from physical interaction, learning in close social contexts, and collaborative embodied learning in general. Embracing the spirit of the rich and perceptive ethnographies so generously shared by this section's authors, I also briefly reflect on a recent experience of my own, my first opportunity for embodied collaborative learning in a pandemic-fatigued world.

Reflections on the experts' accounts

The piece by Kath Bicknell and Kristina Brümmer rotated our worldview by 180 degrees as they cartwheeled us through their experience attending a 'Handstands Foundations' class over a six-week period, with the aim to train the strength, stamina, balance and overall body awareness to achieve a strict handstand. They share their triumphs and tribulations as they chart a rocky, unpredictable and far from linear or smooth learning trajectory from novice to slightly more experienced handstanders. They emphasize the vital role played by failure, regression, near-misses and surprise successes in the learning process (particularly at the early stages), as well as the role(s) played by other participants in the handstand classes (two teachers and twelve learners). Bicknell and Brümmer describe how both expert and novice observations of their own handstand practice helped educate their attention towards proper body alignment, while physical adjustments of the handstander's body by a fellow student or teacher further helped to embed the kind of embodied knowledge required to sustain a well-aligned handstand.

Clear takeaways for me from this piece included not just the value in examining failures and setbacks during complex new physical skill learning among novices, but also how crucial learning companions are for contextualizing and accelerating one's own learning process. If these same classes were offered online during the pandemic, where participants could practice in the comfort of their own (obstacle-cleared) living rooms, it seems obvious that learning would be far slower, patchier and ultimately poorer without the value added by embodied, physically present learning companions sharing the same space with each other as well as instructors. That said, the proliferation of online classes that have emerged as a result of studios and training centres closing their doors, for everything from handstands to tumbling and acrobatics, suggests that at least some aspects of these complex skills can be learned or maintained in a digital context. Specifically, online learning appears to amplify some aspects of the ecology of collaborative, embodied learning (such as discussion of how and where to focus attention, and verbal feedback from an instructor or observer), while other aspects are entirely missing during screen-based learning (such as the trust in an instructor or fellow learner to catch you when you fall).

The value of touch and tactile communication during the learning process was further explored by Downey. In this vignette, we were transported onto a sunny Sydney pool deck, where Downey considers the varieties of collaboration that are essential for building extreme breath-hold discipline, as required by skilled freedivers. These include disrupting habitual breathing patterns, incorporating distraction tasks or instructional nudges into one's practice, leaning on an experienced instructor and/or other learners to scaffold one's own learning and learning to use the physical infrastructure and dive buddies to minimize risk of blackout or obstacles to resurfacing. As Downey elegantly describes, building these skills from the ground up would not be possible without the wealth of knowledge about the human body provided by the medical community, nor would learning and breath-hold development progress as effectively without support from a diving buddy. Downey draws on the classic developmental psychology work by Lev Vygotsky as well as more contemporary work by social anthropologist Jean Lave when he emphasizes that 'learning is most often a social activity, situated interactively in everyday life'.

Similar to the handstand training described by Bicknell and Brümmer, but arguably even more salient due to the life and death nature of the skill being learned, it is difficult to imagine how the type of disciplined breath-hold techniques practised by Downey could be effectively or safely learned *without* other practitioners and/or an expert instructor, all collaboratively learning together. Downey perceptively addresses the seeming paradox of how a skill like freediving, which appears an ideal exemplar of a solo activity, in fact requires deeply collaborative practices to effectively learn. Ultimately, we learn how the skill described in this chapter, which requires individuals to first *unlearn* aspects of the most fundamental motor behaviour of all terrestrial animals, is deeply socially embodied. New learners ignore the value of collaborative learning in this particular context at their peril. Online or screen-based instruction might help build awareness and theoretical knowledge of the skills required and risks encountered in this specific learning context. However, the value added by working with and among others to build this kind of breath-holding skill must be incalculable.

In the penultimate piece, we return to dry land and the feeling of thick gym mats beneath our feet as Susanne Ravn examines the complex antagonistic interactions, where the aim is to win the fight without harming the attacker, that typify aikido. She describes how she becomes part of the aikido ecology – dressing the part, moving the part and socially embodying the part. Her focus on the bodily sensations of being enveloped with stiff cotton fabric, hugged by a fabric belt, feeling an attacker's energy and intentions through a gripped wrist underscore the role of corporeal and physical awareness, first of oneself and one's own body, and then extending to the opponent. She eloquently describes the process of dynamic attunement, where each opponent's attention and sensory awareness is synchronized with the other, and shaped by each individual's skills and abilities (Mingon and Sutton 2021). Ravn's nod to Merleau-Ponty was equally enlightening, in terms of how thinking about how others' bodies helps us to more fully comprehend and engage with the world within one's own body. As Ravn writes, '[t]he body of the other person presents a miraculous prolongation of my intention already on the level of operative intentionality. I sense the intentions of others immediately in their actions, and act in coordinated ways along with the movements initiated.' This way of thinking provides a distinct and thoughtful counterpoint to the focus of experimental psychological literature on joint action (e.g. Sebanz, Bekkering and Knoblich 2006; Sebanz and Knoblich 2021), where much of the focus is on the psychological mechanisms that enable effective action collaboration, such as the analysis and prediction of the specific action parameters of a co-actor's movements. While Ravn touches upon similar themes to those studied by joint action researchers, she is more philosophical and poetic in her consideration of the embodied collaboration, competition and learning that takes place during aikido, suggesting that 'sense-making reaches beyond the here and extends before the now of the two [individuals] practicing together'.

As with learning handstands and breath-holding, Ravn's aikido practice further reinforces, and perhaps even amplifies, how this kind of learning and expertise building centres on the physically embodied practice where bodily knowledge is transmitted, and energy is read and interpreted, through tactile engagement with another practitioner. Far from being based solely on physical aspects of the martial art, the ecology of aikido is further developed via the codes of conduct, techniques, values and rituals of this practice. Gaining a foothold into this practice, and cultivating expertise, will require intentional *and embodied* engagement, and thus close physical contact with another. It is difficult to imagine how the basic building blocks of this practice could be established or consolidated via disembodied, screen-based means.

The final piece, by Roberts and Krueger, takes an exciting and unexpected detour from these grounded, embodied, here-and-now physical skill learning scenarios. Here, the authors consider the status (and legitimacy) of AI systems as creative collaborators in art-making processes. Specifically, they focus on AI's potential for creative agency in music making contexts, shining a spotlight on musician Holly Herndon's collaboration with artificial neural network Spawn during creation of her full-length album, *Proto*. Roberts and Krueger highlight three qualities of musical agency that could feasibly be used to argue against assigning Spawn a role as a full-fledged collaborator: embodiment, emotional expression and autonomy. Across all counts, Spawn and her AI brethren fare poorly (at least at present – although technological advances mean that artificial versions of all three qualities are becoming increasingly sophisticated; see Hortensius et al. 2018). However, Roberts and Krueger argue that it is perhaps more useful to change the narrative about the status or legitimacy of AI's musical agency *per se* to instead consider a *fictionalist* approach to artist–AI collaborations. According to this idea, 'it can be advantageous for an artist or listener to engage in the fictional pretence that there is AI musical agency – ranging from performance and interpretation to full creative composition – even if we accept that this is not literally true'.

Roberts and Krueger's fictionalist proposal does not directly ask what listeners actually think about the origins of an AI in a collaborative artistic role. However, this question fascinates me, as research by my own team and others, suggests that people's beliefs about the autonomy or agency of an AI-imbued agent, or the humanness of an artist, can profoundly shape perceptions and enjoyment of an artwork (or any other stimulus for that matter; e.g. Chamberlain et al. 2018; Cross et al. 2016). Further provocative questions arise when one considers the role played by the human programmer behind an AI algorithm, and their role in establishing the parameters for the AI to learn from, as well as overseeing and fine-tuning the machine learning and training processes. Digging a little deeper into the role of Spawn in Herndon's album, I came across this dazzling quote on her website (emphasis my own):

> You can hear traces of Spawn throughout the album, developed in partnership with long-time collaborator Mathew Dryhurst and ensemble developer Jules LaPlace, and *even eavesdrop on the live training ceremonies conducted in Berlin, in which hundreds of people were gathered to teach Spawn how to identify and reinterpret unfamiliar sounds in group call-and-response singing sessions*; a contemporary update on the religious gathering Holly was raised amongst in her upbringing in East Tennessee.
>
> ('Proto', n.d.)

How utterly fascinating that large groups of people were required to physically come together, to use their voices and bodies to provide the vital human input to train Spawn's algorithms so that this artificial neural network was a suitably creative collaborator. While listeners might willingly (and happily!) engage a fictionalist mindset when considering AI-driven programmes as legitimate creative collaborators, at their core (or at least at Spawn's core) is the rich input of a vast human chorus, with individuals collaborating with each other and the AI to build something akin to musical agency. The questions raised by Roberts and Krueger in their chapter about the legitimacy and status of artificial artistic collaborators are timely and important, as algorithms take on increasingly complex and central decision-making roles in our lives. A clearer understanding and more nuanced appreciation of the human creators who set the wheels in motion for these autonomous artificially intelligent systems to participate in the creative process should spark further spirited debate on the role for digital algorithms in art creation.

Returning to a world of embodied collaborative learning

The physical learning and expertise-building experiences reported by Bicknell and Brümmer, Downey and Ravn, and the AI-training sessions with hundreds of members of the public that were required to breathe life into the artificial musical collaborator profiled by Roberts and Krueger, occurred well before the arrival of a worldwide pandemic that shut most people into their homes for weeks or months on end. While we will undoubtedly see a surge of original research published over the coming months and years documenting learning changes due to a sudden and prolonged shift to disembodied, screen-based learning (where learners can only mix and mingle with other learners and expert instructors from the confines of their square of the Zoom screen), the deeper, more enduring and philosophical implications of pausing embodied collaborative learning and replacing it with proxies will be equally important to understand. Each piece across this section underscores the importance of others' bodies for learning, for improving, for understanding and for creating.

Drawing on my own experience as a practitioner and teacher of contemporary dance, if I wish to teach you how to perform a new dance sequence in an online learning context, I might be able to see that you are holding your shoulders too high even if you appear as a small figure on my laptop screen, or that your timing is slightly off. However, by standing beside you, sharing the same physical space with you, I will have a much better chance of learning that the tension comes from your neck or the base of your spine, just as you might pick up a subtle weight shift that makes timing easier that simply is not discernible in the small digital rendering of an instructor. These sentiments echo observations made by Ravn in her aikido learning journey. Her subtle duet with the elderly Taka-sensei, where his decades of embodied expertise subtly and resolutely undermined her own movements, simply could not happen with anywhere near the same level of corporeal knowing or awareness (if at all), if the practitioners engaged solely via screens.

These insights have been reinforced by recent first-hand experience, when I returned to the dance studio with a group of other dancers, after nearly a full year's absence, to take a breaking course with leading Australian breakdancer b-girl Raygun. After leading us through an athletic warm up and extensive wrist stretching, b-girl Raygun began putting us through our paces by coaxing our bodies through the breakdancing basics, including leg hooks, top rocks, six-step combinations and baby freezes. The twenty or so other workshop participants ranged from complete novices in dance (and breaking) to those with some breaking experience all the way through to professional hip-hop and contemporary dancers and teachers (but for whom breaking was new). As I contorted my body to try to balance all my weight on one ear while resting my ribs on one elbow in order to move into the baby freeze position, my progress was aided by watching other dancers also attempt to negotiate this tricky position, as well as by receiving physical feedback and adjustments from other dancers and b-girl Raygun herself. Sure, one can hold a dance class, an orchestra practice or choral auditions online, but the value and richness of the learning experience, and quality of information transmitted and received simply does not (yet) compare.

The practising of giving and taking control with other practitioners in dance, acrobatics, martial arts and music improvisation, of having conversations with our bodies (not just our heads and upper torsos, floating on a screen, against backgrounds of our chaotic homes and makeshift offices, or perhaps an anodyne workplace template) and physically engaging with other learners has been vital for human learning, creating and thriving for millennia. The beating heart of embodied, contextualized learning is diminished (or at least very much changed) when we are sat behind screens or when pandemic-mandated extreme caution around other living, breathing human beings requires avoidance of physical proximity (let alone contact) with others' bodies. While I am certainly not suggesting we neglect public health advice for the sake of having richer, collaborative embodied learning experiences, I am very much looking forward to a time people, the world over, can safely engage with the kinds of skilful learning activities explored in this section. The countless online dance, music, sport and other digital learning tools that have emerged during this unusual time will undoubtedly reveal those aspects of skill learning that can be learned or refined just as well (or even better) in screen-based compared to in-person contexts. We will also undoubtedly continue to be surprised, disappointed and delighted by technological innovations aimed to emulate or replace human instructors, co-learners and collaborators. However, reflecting on the role of embodiment, and being deeply, physically and completely embedded within a context where learning happens, the chapters in this section, and indeed, across the whole book, reinforce the fact that we were never meant to be isolated, visual creatures, taking in our rich, complex social world with a pair of eyeballs through a phone or computer screen. Returning to Downey's quote about the path to expertise being signposted by the community, my hope is that this unusual time will bring a fuller appreciation that learning can happen with the support of multiple manifestations of community (online and in-person, expert and novice, local and distant), but let us never underestimate what embodied, in-person, and social cooperation brings to complex skill learning and collaboration.

References

Chamberlain, R., C. Mullin, B. Scheerlinck and J. Wagemans (2018), 'Putting the Art in Artificial: Aesthetic Responses to Computer-Generated Art', *Psychology of Aesthetics, Creativity, and the Arts* 12 (2): 177–92.

Cross, E.S., R. Ramsey, R. Liepelt, W. Prinz and A. F. de C. Hamilton (2016), 'The shaping of social perception by stimulus and knowledge cues to human animacy', *Philosophical Transactions of the Royal Society B: Biological Sciences*, 371 (1686), 20150075.

Hortensius, R, F. Hekele and E. S. Cross (2018), 'The Perception of Emotion in Artificial Agents,' *IEEE Transactions on Cognitive and Developmental Systems* 10 (4): 852–64. doi: 10.1109/TCDS.2018.2826921.

Mingon, M. and J. Sutton (2021), 'Why Robots Can't haka: Skilled Performance and Embodied Knowledge in the Māori haka', *Synthese*. https://doi.org/10.1007/s11229-020-02981-w

Proto (n.d.), *Holly Herndon*. Available online: http://www.hollyherndon.com/proto (accessed 21 June 2021).

Sebanz N, H. Bekkering and G. Knoblich (2006), 'Joint Action: Bodies and Minds Moving Together', *Trends in Cognitive Sciences* 10 (2): 70–6.

Sebanz N. and G. Knoblich (2021), 'Progress in Joint Action Research', *Current Directions in Psychological Science* 30 (2): 138–43.

PART THREE

Symmetry and synergy in embodied coordination

8

Symmetries of social performance-environment systems

Rachel W. Kallen, Margaret Catherine Macpherson, Lynden K. Miles and Michael J. Richardson

Many everyday behaviours unfold in social settings in ways that demand coordination with the actions of others. Ranging from relatively simple exchanges when people avoid one another on a crowded sidewalk or move a piece of heavy furniture together, to more complex interactions such as ensemble performances or competitive sports, coordination between individuals underlies effective social behaviour. Given the multiplicative forms of social interaction, attempting to understand the organization and patterning of social coordination from a unified theoretical perspective might seem unachievable. However, research over the last several decades within the psychological and cognitive sciences offers converging evidence that the patterning of social coordination is defined by the interrelated physical, informational and sociocultural dynamics of *performance-environment systems* (Hutchins 2010; Turvey 2007). Consistent with an embodied-embedded and *Ecological Dynamics* approach to human behaviour

(Araújo et al. 2006; Kelso 1995; Turvey 2018; Warren 2006) (Figure 7a), and the contemporary complex systems approach to understanding biological and social phenomena (Miller and Page 2007), the provocative implication is that the structure of social interaction could be understood using lawful (i.e. dynamical), context-sensitive principles.

Identifying what these principles are, however, remains the key barrier to developing a comprehensive account of complex social (and individual) performance. Indeed, research exploring the emergence of complex behaviour often falls into the trap of either: (i) reducing phenomena to overly parsimonious mathematical formalisms (i.e. computational rules); or (ii) merely describing phenomena using the terminology of complex systems without any analytical rigour. What is needed, therefore, is a theoretical (meta-)language that is not only consistent with a complex performance-environment systems approach, but can also capture the complementary nature of extant accounts of self-organized social and individual experience (Pattee 1978). Here, we propose that the theoretical and formal concepts of symmetry and symmetry breaking might provide that language (also see Richardson and Kallen 2015). In this chapter, we attempt to illustrate how the formal (mathematical) and theoretical principles of symmetry and symmetry breaking underlie both the potential and observed structural organization and patterning of a wide array of social performance phenomena, including motor coordination and behavioural synchrony, improvisation and creativity, sport and competitive interaction, and social and cultural systems.

Symmetry and symmetry breaking

The term symmetry refers to the invariance ('sameness') of something given some form of transformation (change). For instance, the approximate bilateral symmetry of the human face is due to the invariance of form when reflected about the midline axis. Similarly, the symmetry of geometric shapes corresponds to the fact that such shapes look the same when rotated or reflected in certain ways (Figure 7c). Importantly, the principle of symmetry can be generalized beyond the structure of geometric objects and, as a scientific principle, is employed to understand the ordered patterning of everything from the simple arrangement of objects and system properties (Figure 7c), to abstract mathematical functions and the fundamental laws of nature and the universe (Stewart and Golubitsky 1992). The principle of symmetry even forms the basis of the scientific method, in that the reliability, or invariance (symmetry) of an experimental result is determined via transformations in time and space (e.g. do results obtained by lab-A in 2019 replicate using the same experimental method by lab-B in 2021?).

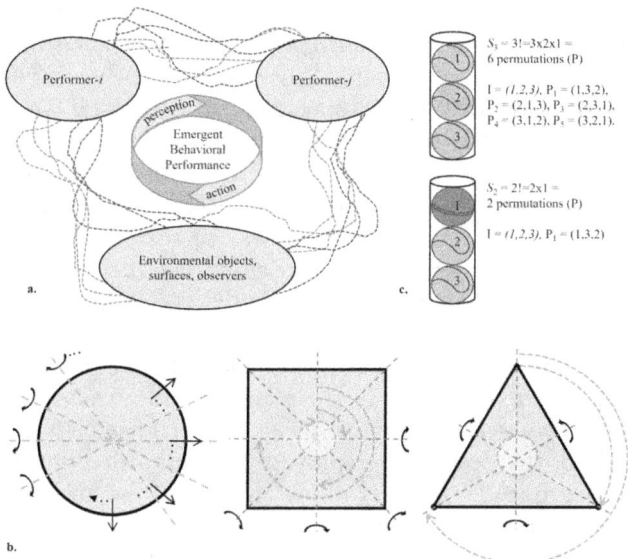

FIGURE 7 *(a) Ecological Dynamics approach to performance-environment systems. Behavioural performance emerges from embedded perception–action cycles and agent–environment interactions. (b) A circle has an infinite number of rotations and reflections captured by orthogonal group O (2). A square has eight symmetries, four rotations = {0°/360°, 90°, 180°, 270°} and four reflections, captured by the dihedral group, D_4. An equilateral triangle has six discrete symmetries, three rotations = {0°/360°, 120°, 240°} and three reflections, captured by the dihedral group, D_3. (c) (top) An illustration of how the symmetry group S_3 can capture the spatial transformations (permutations) of three identical objects; tennis balls stacked in a cylinder. (bottom) An illustration of how replacing one tennis ball with a non-identical cricket ball reduces the symmetry group of the three objects to the permutation group S_2, which is a subgroup of S_3. 'I' refers to the identify (do nothing) transformation.* © Michael Richardson.

Group theory

Symmetry can be quantified, in that some object or system can have more or less symmetry than another object or system. Consider the geometry of a circle, square and an equilateral triangle (Figure 7b). A circle has a larger set (group) of symmetries than a square, which in turn has a larger group of symmetries than the triangle. This formal quantification of symmetry is achieved using the abstract algebra of Group Theory where the symmetry group of an object or system corresponds to the closed set of transformations that leave the object or system unchanged. While detailed discussion of the mathematics of Group Theory is beyond the scope of this chapter (see Richardson and Kallen 2015; Rosen 1985; Weyl 1952), the group concept is required to explicate how symmetry can be used to understand several of the social performance phenomena we explore. Of significance here is that defining the symmetries of something by means of a group allows one to formally determine the dynamical or structural correspondence between different systems (e.g. how behaviour of seemingly *different* systems results from the *same* system of constraint). Moreover, the subgroups of a symmetry group can be used to predict what potential patterns or structures of behavioural order are possible when changes in a system's symmetry occur (i.e. when symmetry is broken; Figure 7b). Indeed, the power of employing principles of symmetry and symmetry groups is that they allow one to develop an explanation of behavioural phenomena that is both abstract (conceptual/descriptive) and formal (mathematical/predicative).

Symmetry breaking

Although Group Theory provides a way of formally defining the structural organization of a system, key to understanding why certain patterns of behaviour emerge over time is understanding the role that symmetry breaking plays in the creation of behavioural order. In short, paraphrasing Pierre Curie (1894), *it is asymmetry [broken symmetry] that creates phenomena* (cited in Park and Turvey 2008).

As an intuitive example of how symmetry breaking is fundamental to the emergence of behavioural order, consider being seated at an immaculately set circular dining table, with dinnerware positioned symmetrically around the table. Although pleasing to the eye, such arrangements create uncertainty about whose glass is whose. If unfamiliar with social convention, the solution to the dilemma of 'whose glass is whose' is ill-defined; everyone seated at the table exists at the precipice of a right- or left-glass state. This dilemma is easily solved, however, as soon as one individual grasps a glass. If that individual chooses a glass to the right, the collective order of the group immediately collapses to a right-glass state, with every other individual in the group choosing the glass to the right (including those not yet seated at the table).

This example highlights several important points. First, *the order of behaviour emerges from symmetry-breaking events*. Here, order refers to the collective organization or patterning of behaviour. Thus, for the 'whose glass is whose' example, selection of an individual glass breaks the symmetry of the table, thereby defining the pattern of coordination that individuals engage in. Second, *more symmetry is synonymous with less order* and, conversely, *less symmetry is synonymous with more order*. Indeed, before any glass is chosen, the symmetric state of the table layout results in a lack of behavioural organization; however, once the symmetry is broken a stable pattern of spatiotemporal organization emerges. Finally, it is important to appreciate that *symmetry-breaking events do not simply correspond to the destruction of symmetry, but rather a redistribution of symmetry*. Exemplified in the above example, the initial symmetry brought about by indecision gives way to the new symmetry of all individuals using the glass to their right. Moreover, the terms symmetry and asymmetry are relational concepts, one implies the other, in that for something to have symmetry there must be an asymmetry by which that symmetry is gauged.

Social motor coordination

As an example of how the principles of symmetry shape the structure of social performance, consider the behavioural coordination that occurs between the rhythmic movements of co-present individuals. As you may have experienced when walking and talking with a friend, the rhythmic actions of human co-actors often become synchronized. Extensive research has demonstrated that such behavioural synchrony can be understood and explained by the dynamics of coupled oscillators (Schmidt et al. 1990). Of relevance is that the formal principles of symmetry define how rhythmic movements that have the same natural frequency become synchronized when the coupling between the movements is sufficiently strong (Schmidt and Richardson 2008). For instance, if two individuals sitting side by side in identical rocking chairs are looking at one another, even intermittently, their movements tend to become synchronized over time (Richardson et al. 2007). That is, at any point in time, the spatiotemporal position of both chairs will be the same and, hence, the movements will be invariant with respect to inversion (i.e. symmetrical); the patterning of the coordination would look the same if the individuals swapped chairs or if we observed their behaviour in a mirror.

In contrast, consider what would happen if we induced a break in the symmetry of the system by changing the size and/or weight of one rocking chair. For example, in the situation where an adult is rocking in a large chair and a child is rocking in a small chair, the natural rocking frequency of the adult's chair would be much slower than the child's. In turn, the behavioural synchronization that occurs would entail a phase lead/lag, with the adult's movements synchronized slightly behind the child's. In other words, the induced asymmetry in the natural frequency of movement results in a corresponding asymmetry in the pattern of coordination that emerges, with the resulting deviation from perfect synchronization specifying that system 1 (slow, adult) is different from system 2 (fast, child).

Clearly, increases in behavioural order resulting from changing the natural frequencies of rocking chairs, or any rhythmic movement, are rather minimal. However, it is important to appreciate that larger changes in the ratio of movement frequencies can bring about the kinds of complex polyrhythmic behavioural patterns inherent to most forms of music and dance. It also underlies the 2:1 or 3:1 step-skip patterns a small child produces when trying to walk alongside a 'longer-legged' and faster adult. Moreover, the rocking chair example highlights how perfect 1:1 behavioural synchronization is a 'special' case of coordination, in that it only occurs when the movement properties are the same (symmetric), and it is spontaneous or induced asymmetries in system properties, be they physical, informational, cognitive, social or otherwise, that create the possibility for more complex patterns of non-synchronous behavioural coordination to emerge.

Finally, note that perfectly symmetric or synchronous coordination is often degenerative and non-functional, in that its realization can hinder rather than promote task-oriented behaviour (Kijima et al. 2017). Consider walking down a pathway while another individual is walking directly towards you (an everyday 'game of chicken'). One often finds oneself caught in a non-functional synchronous dance with the other individual, both moving back and forth in the same direction at the same time, preventing either from moving on. In such cases, you or the other individual must stop to break the symmetry of impassability. Alternatively, to prevent the uncomfortable situation from arising altogether, one person must 'move' first (or nonverbally indicate via nod or gaze), to define the asymmetry of either 'you go to the right, I go to the left', or perhaps even the more socially differential situation of, 'you go straight, I will go around' (Meerhoff et al. 2018; Richardson et al. 2015).

Social improvisation and creativity

In his renowned book, '*Symmetry*' (1952), Hermann Weyl famously wrote that 'beauty is bound up with symmetry'. Indeed, it is well known that humans have an affective pull towards symmetric patterns with symmetry playing a foundational role in the aesthetics of the visual and performing arts, and architecture throughout the ages. However, more symmetry does not equal 'more beautiful'. On the contrary, for something to have a pattern that can be observed and appreciated, a corresponding asymmetry must exist to define what is symmetric. To illustrate, imagine a perfectly white painting on a perfectly white canvas, hung on a perfectly white wall, inside a perfectly white room, lit with perfect white light. In this perfectly symmetric space, one would not be able to see the painting at all, let alone appreciate the painting as a distinct piece of art. However, given the slightest change to any one element in the room, the colour of wall, or the light, all of a sudden, a separation between painting and wall would occur, and one could identify the painting as being its own structure. A few strokes of coloured paint and one might also have a more interesting painting, where the 'beauty' of the painting's form, the artist's creativity, is fundamentally tied to a dualistic balance of symmetry and asymmetry. In this sense, the aesthetic appeal of something is not 'how' symmetric or asymmetric it is, but the symmetry–asymmetry relation being expressed.

To return to social performance, consider the symmetries that must exist for a group of musicians playing a piece of music together. The style of music being performed (e.g. jazz, classical) is already defined by the symmetry of the musical style; each musical style has a genre specific structure that is perceived to be the same over different pieces of music. A certain degree of symmetry must also exist during the performance, whereby each musician must follow the same referent and play in time with one another. However, it is the asymmetries of different musical instruments, musicians, octaves, syncopation and beat-to-beat timing that both defines the symmetry of the performance and the piece of music being performed, while simultaneously allowing the performance to be unique and expressive. In stark contrast, a group of people playing the same note, on the same instrument, at the same time, may have perfect symmetry, but no musical form, nor any groove, emotion or feel. It is in this way that symmetry breaking is foundational to creativity and improvised social performance. To paraphrase Donnini (1986), new performance structures are reached through a reconstruction of a pre-existing referent. The introduction of a novel difference causes an asymmetry that makes headway towards another asymmetry and so on, in a manner producing, in the end, a new symmetrical order (Donnini 1986, 436).

Walton and colleagues (Walton et al. 2018) have referred to such musical creativity and improvised decision making as the 'making of waves', with the waves that emerge constrained by the degree of symmetric structure imposed by the musical context (e.g. score, differences in expertise, musician's personality). Importantly, 'new waves' of musical improvisation must still adhere to the group transformations that define musical composition. For example, the closed set of transpositions and reflections that change the register (i.e. octave, pitch) or invert a melodic pattern, respectively, while ensuring tonal and melodic harmony (Cohn 1997). Indeed, musical theory is a theory of symmetry transformations, with Group Theoretic principles explicating the sequencing and progressive structure of Beethoven's symphonies through to Ozzy Osbourne's rock anthems (Capuzzo 2004).

Schaffer (2014, 2019) has detailed how symmetry and symmetry groups also define the body poses and movement patterns observed in individual, interpersonal and group dance, along with other movement arts (e.g. gymnastics, skydiving). Using symmetry groups that define transformations in three-dimensional space, Schaffer illustrated (a) how the symmetry transformations of rotation and reflection map the body and limb postures and movements of dancers within their body and across two or more bodies; (b) how symmetry transformations in time define how dancers transform one structural configuration or sequence of dance moves into an isomorphic configuration or sequence later in time; and (c) how dancers embed the nature of their performance, by structuring the symmetry of their movements in relation to the perceived structure (affordances) of the spatial and auditory environment they are performing in (e.g. shape or symmetry of the stage/set, lighting, music).

Sport and competitive interaction

Patterns of behavioural performance in sport and competitive interaction can also be understood using the formal principles of symmetry and symmetry breaking. Yokoyama and Yamamoto (2011) demonstrated how the patterns of behavioural coordination observed between three teammates passing a soccer ball to prevent interception from an opponent are constrained by the lower order subgroups of the symmetry group that defines the possible patterns of coordination between three-coupled oscillators (i.e. $D_3 \times S_1$); namely: (1) all teammates rotate body positions 120 degrees out of phase with one another (a rotating wave configuration); (2) two teammates rotate their body position in the same (in-phase) relative direction as one another, while the third teammate rotates in an opposite (antiphase) or independent direction; and (3) two teammates rotate body position in opposite (antiphase) relative directions, while the third teammate remains still.

Importantly, the Group Theoretic approach employed by Yokoyama and Yamamoto was derived from the Group Theoretic understanding of coupled oscillatory systems termed Symmetric Hopf-Bifurcation theory (Collins and Stewart 1994). In short, this approach defines the patterns of behavioural coordination that are possible across any number of coupled oscillators (2, 3, ... to n), and is able to explain the patterns of interlimb coordination observed across the animal kingdom, from the synchronous states of social motor coordination defined above to the gait patterns of bipeds, quadrupeds and hexapods. Thus, the work of Yokoyama and Yamamoto not only highlights the explanatory power of symmetry groups, but the degree to which Group Theoretic explanations can reveal the dynamic similitude of seemingly different behavioural phenomena.

As well as the order of behavioural coordination within a team, symmetry defines the patterns of coordination between teams, most notably in offense–defence arrangements. Consider the organization of players in the backline of opposing rugby teams. The goal of the defending team is to match, player to player, the arrangement of the players on offense, ensuring that the offensive team cannot break the line and score a try. In contrast, the goal of the offensive team is to break the symmetry of player-to-player marking, creating the opportunity (i.e. a 'gap') for one player to penetrate the defensive line and score a try (Vilar et al. 2012). Over the last two decades, Araújo, Davids and colleagues have articulated the role of symmetry and symmetry breaking across a range of team sports, providing rich theoretical and empirical evidence for how the emergence, maintenance and destruction of patterns of offensive and defensive play is based on the functional interactions between players and the dynamics of the performance-environment system (Araújo et al. 2006; Davids et al. 2013). Consistent with the ecological dynamics approach, they highlight how dynamic patterns of game play emerge from critical fluctuations in the balance of the symmetry-preserving and symmetry-breaking actions of opposing players, with such fluctuations often resulting in the sudden reorganization of offensive/defensive game play at both short (seconds/minutes) and long (match/series) time scales.

This reciprocal tension between maintaining and breaking symmetry is not simply inherent to team sports but underlies all competitive performance contexts. For example, this tension is readily observable in combat sports such as boxing, martial arts and sword fighting (Caron et al. 2017) as well as racket sports, such as tennis and badminton (McGarry et al. 2002). Analogous to the 'whose glass is whose' and 'interpersonal chicken' examples, opponents in dyadic sports often begin to engage with each other in a symmetric state (e.g. synchronously moving back and forth; rallying a ball back and forth), poised in behavioural pattern that entails multiple possibilities (some that might be anticipated and some that might not). This symmetry, however, is broken when one player gains the advantage (Kimmel and Rogler 2018) with the asymmetric dominance of one player over another reflected in the behavioural outcomes of winner and loser.

At a broader level, even the rules of sport are defined around the principles of symmetry. Equity (symmetry) to ensure safe and fair competition (the same number of players on opposing teams), and induced inequity (asymmetry) when the rules are broken (penalties to create asymmetries in possession or position; red cards inducing an asymmetry to disadvantage offending player's team) are the cornerstones of organized (rule-bound) competitive interactions.

Social systems

To this point, the examples of symmetry and symmetry breaking discussed predominantly consider the performance of coordinated behaviour in relatively task-driven contexts. Here, the symmetry-breaking events described typically lead to patterns of behaviour that support functional task-relevant outcomes (e.g. avoiding collisions, improvising a duet, beating a defender). When conceptualized more broadly, the performance of everyday behaviour also unfolds within a wider sociocultural context whereby a host of additional factors (e.g. attitudes, social norms, group membership) come into play that constrain patterns of interaction. Moreover, daily experience attests that such factors can promote both functional, positive (e.g. rapport and affiliation) and dysfunctional, negative (e.g. stereotyping and prejudice) social outcomes.

Thus, to be a generalized framework, the account of symmetry breaking proposed here must also extend to explanations of social psychological and cultural phenomena when attempting to understand the coordination of interpersonal behaviour. Returning to a central theme of the chapter: *(a)symmetry of (social) behaviour is dependent on the (a)symmetry in the (social) factors that bring about that behaviour*. Indeed, the symmetry (stability over transformations in time) of social and cultural systems is the result of broken symmetry. That is, the symmetry of individual, group and cultural identity (e.g. what separates me from you, us from them) is dependent on the presence or induction of asymmetry. What is required, therefore, is to map the relationships between (a)symmetries in socially relevant factors and (a)symmetries in socially relevant behaviour.

As a backdrop for conceptualizing the role of symmetry at the social level, it is important to note a common theme in the empirical literature – coordinated behaviour has affiliative properties (Mogan et al. 2017). Broadly speaking, we align our behaviour with those we like and conversely, we like those we align our behaviour with. In this way symmetry and symmetry breaking are thought to show a mutual interdependence at the social and motor levels. Support for this proposition can be found in experimental work in which symmetry breaking events manipulated at the social level have been shown to impact the patterning of interpersonal motor behaviour. For instance, Miles et al. (2010) had half of their participants wait an extended period of time for a 'late' confederate. In the sociocultural context of the experiment, the asymmetry in time between the participant and confederate brings an additional social meaning – the confederate's behaviour can be interpreted as rude (a view which, upon arrival, she intentionally did nothing to dispel). During a subsequent interaction, what unfolded was a reduction in both social motor coordination and rapport between the confederate and participant, whereby the asymmetry in social etiquette (i.e. differences in punctuality) brought about a concomitant asymmetry in coordinated behaviour and affiliation. Patterns of social behaviour, in terms of differences in rapport, emerged from a symmetry breaking event and, in contrast to the examples considered to this point, resulted in a *negative* outcome – a breakdown in rapport.

Beyond transgressions of social norms, at the interpersonal level symmetries and asymmetries can emerge over longer timescales as a function of socially relevant individual differences. In early work, Schmidt et al. (1994) demonstrated that pairs of participants matched for social competence (i.e. both individuals were either high or low in social competence) produced less stable patterns of interpersonal coordination when compared to pairs with differing (asymmetrical) social competence. Here, broken symmetry supported stable patterns of coordination. On the other hand, asymmetries in other individual difference factors, in particular psychopathology, have been noted to disrupt the stability of interpersonal coordination (Macpherson et al. 2020). Across a range of disorders of social functioning, including autism spectrum disorder (ASD) (Fitzpatrick et al. 2017; Hussein et al. 2017), schizophrenia and social anxiety disorder (SAD) (Varlet et al. 2012, 2014), asymmetric dyads (i.e. one individual with a diagnosis and one healthy control) exhibit atypical patterns of interpersonal synchrony. These examples underscore an important point – asymmetries at socially relevant levels of explanation can lead to new patterns of socially relevant behaviour. The order of collective behaviour, functional or dysfunctional, emerges from symmetry-breaking events.

The social effects of symmetry and symmetry breaking are also evident at community or population-based levels of analysis. Schelling's ground-breaking work (Schelling 1971) outlined an agent-based model that attempted to simulate patterns of ethnic segregation observed in many major cosmopolitan cities. Starting with two kinds of agents randomly distributed across a two-dimensional 'world', Schelling's model implemented a simple rule: on each turn agents should move to a new (random) location unless a certain number of their immediate neighbours were of the same kind. This rule mimicked a typical in-group preference widely documented in the social psychological literature (Brewer 1979), and enabled manipulation of the 'tolerance' of the agents to diversity. What emerged was a threshold at which segregation of a formerly integrated population of agents occurred spontaneously. Specifically, if at least one-third of an agent's neighbours were the same, over time, agents of the same kind clustered together. A small asymmetry in individual preferences for a type of neighbour resulted in a highly ordered (and stable), yet undesirable and dysfunctional, pattern at the population level. Indeed, recent work has indicated striking similarities between Schelling's model and residential patterns in contemporary urban dwelling – cities where survey data reveal in-group ethnic preferences typically show high levels of ethnic segregation (Clark and Fossett 2008), a pattern of living associated with high levels of inequality and social disadvantage (Massey and Fischer 2000).

Finally, we conclude with a more anecdotal account of symmetry breaking triggering positive societal change. In 2016 Colin Kaepernick, an American football player, repeatedly 'took a knee' during the national anthem prior to NFL games in protest of systemic racial inequality in the U.S. An explicit and highly visible instance of symmetry breaking, kneeling during the anthem, created information by drawing attention to a contentious social issue and fostered heightened societal awareness. When understood in the sociocultural context in which it occurred, the performance of this otherwise innocuous act created an asymmetry that contributed to the emergence of new patterns of coordinated behaviour – a protest movement intended to generate a new order of positive societal change.

Conclusion

Our aim was to briefly outline a theory of social behaviour and performance grounded in formal principles of symmetry. In doing so, we have touched upon how these symmetry principles and groups can both account for and predict the patterning of social performance across multiple contexts and levels of analysis. Fundamental to this approach is the notion that symmetry-breaking events create opportunities for stable states of action and interaction specific to the context of that system. The explanatory power of this approach comes from its generality – any system across any scale is subject to the same organizing principles – the challenge that remains, therefore, is to continue to validate the predictive power of the proposed approach across a growing body of examples.[1]

Note

1 This work was supported by an Australian Research Council Future Fellowship (FT180100447) awarded to Michael Richardson.

References

Araújo, D., K. Davids and R. Hristovski (2006), 'The Ecological Dynamics of Decision Making in Sport', *Psychology of Sport and Exercise* 7 (6): 653–76.

Brewer, M. B. (1979), 'In-group Bias in the Minimal Intergroup Situation: A Cognitive-Motivational Analysis', *Psychological Bulletin* 86 (2): 307–324.

Capuzzo, G. (2004), 'Neo-Riemannian Theory and the Analysis of Pop-Rock Music', *Music Theory Spectrum* 26 (2): 177–200.

Caron, R. R., C. A. Coey, A. N. Dhaim and R. C. Schmidt (2017), 'Investigating the Social Behavioral Dynamics and Differentiation of Skill in a Martial Arts Technique', *Human Movement Science* 54: 253–66.

Clark, W. A. V. and M. Fossett (2008), 'Understanding the Social Context of the Schelling Segregation Model', *Proceedings of the National Academy of Sciences* 105 (11): 4109–14.

Cohn, R. (1997), 'Neo-Riemannian Operations, Parsimonious Trichords, and Their 'Tonnetz' Representations', *Journal of Music Theory* 41 (1): 1–66.

Collins, J. J., and I. Stewart (1994), 'A Group-Theoretic Approach to Rings of Coupled Biological Oscillators', *Biological Cybernetics* 71 (2): 95–103.

Curie, P. (1894), 'Sur la symmétrie des phénomènes physiques: symmétrie d'un champ électrique et d'un champ magnétique', *Journal de Physique* 3: 393–415.

Davids, K., D. Araújo, L. Vilar, I. Renshaw and R. Pinder (2013), 'An Ecological Dynamics Approach to Skill Acquisition: Implications for Development of Talent in Sport', *Talent Development and Excellence* 5 (1): 21–34.

Donnini, R. (1986), 'The Visualization of Music: Symmetry and Asymmetry', *Computers & Mathematics with Applications* 12 (1–2): 435–63.

Fitzpatrick, P., V. Romero, J. L. Amaral, A. Duncan, H. Barnard, M. J. Richardson, and R. C. Schmidt (2017), 'Evaluating the Importance of Social Motor Synchronization and Motor Skill for Understanding Autism', *Autism Research* 10 (10): 1687–99.

Hussein, A., M. M. Gaber, E. Elyan and C. Jayne (2017), 'Imitation Learning: A Survey of Learning Methods', *ACM Computing Surveys* 50 (2): 1–35.

Hutchins, E. (2010), 'Cognitive Ecology', *Topics in Cognitive Science* 2 (4): 705–15.

Kelso, J. A. S. (1995), *Dynamic Patterns: The Self-Organization of Brain and Behavior*. Cambridge, MA: MIT Press.

Kijima, A., H. Shima, M. Okumura, Y. Yamamoto and M. J. Richardson (2017), 'Effects of Agent-Environment Symmetry on the Coordination Dynamics of Triadic Jumping', *Frontiers in Psychology* 8: 3.

Kimmel, M., and C. R. Rogler (2018), 'Affordances in Interaction: The Case of Aikido', *Ecological Psychology* 30 (3): 195–223.

Macpherson, M. C., D. Marie, S. Schön and L. K. Miles (2020), 'Evaluating the Interplay between Subclinical Levels of Mental Health Symptoms and Coordination Dynamics', *British Journal of Psychology* 111 (4): 782–804.

Massey, D. S., and M. J. Fischer (2000), 'How Segregation Concentrates Poverty', *Ethnic and Racial Studies* 23 (4): 670–91.

McGarry, T., D. I. Anderson, S. A. Wallace, M. D. Hughes and I. M. Franks (2002), 'Sport Competition as a Dynamical Self-Organizing System', *Journal of Sports Sciences* 20 (10): 771–81.

Meerhoff, L. A., J. Pettré, S. D. Lynch, A. Crétual and A. H. Olivier (2018), 'Collision Avoidance with Multiple Walkers: Sequential or simultaneous interactions?', *Frontiers in Psychology* 9: 2354.

Miles, L. K., J. L. Griffiths, M. J. Richardson and C. N. Macrae (2010), 'Too late to Coordinate: Contextual Influences on Behavioral Synchrony', *European Journal of Social Psychology* 40 (1): 52–60.

Miller, J. H. and S. E. Page (2007), *Complex Adaptive Systems: An Introduction to Computational Models of Social Life*. Princeton, NJ: Princeton University Press.

Mogan, R., R. Fischer and J. A. Bulbulia (2017), 'To Be in Synchrony or Not? A meta-Analysis of Synchrony's Effects on Behavior, Perception, Cognition and Affect', *Journal of Experimental Social Psychology* 72: 13–20.

Park, H., and M. T. Turvey (2008), 'Imperfect Symmetry and the Elementary Coordination Law', In A. Fuchs and V.K. Jirsa (eds), *Coordination: Neural, Behavioral and Social Dynamics*, 3–25, Berlin: Springer.

Pattee, H. H. (1978), 'The Complementarity Principle in Biological and Social Structures', *Journal of Social and Biological Systems* 1 (2): 191–200.

Richardson, M. J, K. L. Marsh, R. W. Isenhower, J. R. L. Goodman and R. C. Schmidt (2007), 'Rocking Together: Dynamics of Intentional and Unitentional Interpersonal Coordination', *Human Movement Science* 26 (6): 867–91.

Richardson, M. J., S. J. Harrison, R. W. Kallen, A. Walton, B. A. Eiler, E. Saltzman, R. C. Schmidt and J. T. Enns, (2015), 'Self-Organized Complementary Joint Action: Behavioral Dynamics of an Interpersonal Collision-Avoidance Task', *Journal of Experimental Psychology: Human Perception and Performance* 41 (3): 665–79.

Richardson, M. J. and R. W. Kallen (2015), 'Symmetry-Breaking and the Contextual Emergence of Human Multiagent Coordination and Social Activity', In E. Dzhafarov, S. Jordan, R. Zhang and V. Cervantes (eds), *Contextuality from Quantum Physics to Psychology*, 229–86, Singapore: World Scientific Publishing Co. Pte. Ltd.

Rosen, M. R. (1985), 'Cellular Electrophysiology of Digitalis Toxicity', *Journal of the American College of Cardiology* 5 (5): 22A–34A.
Schaffer, K. (2014), 'Dancing Deformations', In *Proceedings of Bridges 2014: Mathematics, Art, Music, Architecture, Culture*, 253–60. Phoenix, AZ: Tessellations Publishing.
Schaffer, K. (2019), 'Three-Dimensional Symmetries in Dance and Other Movement Arts', In *Proceedings of Bridges 2019: Mathematics, Art, Music, Architecture, Education*, 247–54. Phoenix, AZ: Tessellations Publishing.
Schelling, T. C. (1971), 'Dynamic Models of Segregation', *The Journal of Mathematical Sociology* 1 (2): 143–86.
Schmidt, R. C., C. Carello and M. T. Turvey (1990), 'Phase Transitions and Critical Fluctuations in the Visual Coordination of Rhythmic Movements between People', *Journal of Experimental Psychology: Human Perception and Performance* 16 (2): 227–47.
Schmidt, R. C., N. Christianson, C. Carello and R. Baron (1994), 'Effects of Social and Physical Variables on between-Person Visual Coordination', *Ecological Psychology* 6 (3): 159–83.
Schmidt, R. C. and M. J. Richardson (2008), 'Dynamics of Interpersonal Coordination', In A. Fuchs and V.K. Jirsa (eds), *Coordination: Neural, Behavioral and Social Dynamics*, 281–308, Berlin: Springer.
Stewart, I. and M. Golubitsky (1992), *Fearful Symmetry: Is God a Geometer?* Oxford: Blackwell.
Turvey, M. T. (2007), 'Action and Perception at the Level of Synergies', *Human Movement Science* 26 (4): 657–97.
Turvey, M. T. (2018), 'Lectures on Perception', In *Lectures on Perception*. New York, NY: Routledge.
Varlet, M., L. Marin, D. Capdevielle, J. Del-Monte, R. C. Schmidt, R. N. Salesse, J-P Boulenger, B. G. Bardy and S. Raffard (2014), 'Difficulty Leading Interpersonal Coordination: Towards an Embodied Signature of Social Anxiety Disorder', *Frontiers in Behavioral Neuroscience* 8: 29.
Varlet, M., L. Marin, S. Raffard, R. C. Schmidt, D. Capdevielle, J-P Boulenger, J. Del-Monte, B. G. Bardy and L. Fontenelle (2012), 'Impairments of Social Motor Coordination in schizophrenia', *PLoS ONE* 7 (1): e29772.
Vilar, L., D. Araújo, K. Davids and C. Button (2012), 'The Role of Ecological Dynamics in Analysing Performance in Team Sports', *Sports Medicine* 42 (1): 1–10.
Walton, A. E., A. Washburn, P. Langland-Hassan, A. Chemero, H. Kloos and M. J. Richardson (2018), 'Creating Time: Social Collaboration in Music Improvisation', *Topics in Cognitive Science* 10 (1): 95-119.
Warren, W. H. (2006), 'The Dynamics of Perception and Action', *Psychological Review* 113 (2): 358–89.

Weyl, H. (1952), *Symmetry*. Princeton, NJ: Princeton University Press.
Yokoyama, K., and Y. Yamamoto (2011), 'Three People Can Synchronize as Coupled Oscillators during Sports Activities', *PLoS Computational Biology* 7 (10): e1002181.

9

Sing's trap: Staging low-commitment strategizing in muay thai

Sara Kim Hjortborg

On stage, at the YOKKAO 40 fight show in Sydney, Singpayak (or Sing), a Thai fighter from PTJ muay thai gym in Sydney fought Lloyd Dean, an Australian muay thai champion from Perth known as 'the Nightmare'.

The fight, lasting five rounds of three minutes each, was a prize fight under full muay thai *rules*: minimal protection limited to the groin and light-weight gloves and legal techniques that included punches, elbows, knees, kicks, sweeps and throws.

During the fight, Sing confronted Lloyd with a casual ease: his body relaxed and slightly slumped, his gaze focussed yet tending towards boredom and his arms swinging nonchalantly at his side when just out of reach from his adversary. Lloyd on the other hand started off slightly tense: his arms held aversively in front of him, his demeanour slightly agitated, trying to defend himself from assaults in his attempts to push in and start his own game.

In the second round of the fight Sing landed a well-executed elbow strike to Lloyd's forehead, inflicting a severe cut that almost ended the fight early by technical knockout.[1] When Sing later commented on this moment and his final win, he did not focus on power nor physique. His understanding of what had happened was based on *careful attention and forward planning which he attributed to his ability to 'read' his opponent and 'think ahead'.*

Expert muay thai fighters are required to perform cognitively in the ring. Competing in this particularly brutal and demanding form of combat demands cunning tactics and skilled strategic decision-making, especially in a well-matched high-level fight. Some researchers in skilled performance propose 'strategy' as a central constituent to meet task demands (MacMahon and McPherson 2009). But do fighters really 'strategize'? If so, how, in practice and at high speed? How can a 'strategy' be mapped onto unpredictable, unstable and insecure futures? Or are fighters' actions more directly elicited through skilful environmental attunement and regulation?

In this chapter, I explore a pinnacle moment in a muay thai fight to examine dynamic decision-making processes in a precarious dyadic setting. I map three stages of the decision-making process that drive the development of low-commitment strategies. I call these *diagnosing, probing* and *choosing*. These sequential yet overlapping stages – diagnosing and probing for example usually operate together and in iterative loops – each create constraints, actively facilitate control and nudge the interaction towards self-enabling ends. By low-commitment strategies, I mean to suggest 'directives' rather than actual formulas or rigid 'cut-in-stone' plans to guide action behaviour (Engel 2010; Kimmel and Rogler 2019, 244). If in contrast a high-commitment strategy designates a more inflexible and overly cognitivist account of planning, then, a low-commitment strategy more appropriately captures some form of (thoughtful) commitment that is less specific than a 'plan' (in the conventional sense), yet constrains options and channels action possibilities towards particular outcomes as interaction unfolds (Kimmel and Rogler 2019; van Dijk and Rietveld 2021).

Precarious strategizing in fight interactions operates at the individual level but is highly sensitive to, and in constant process with the environment, including the opponent. A fighting couple must continuously co-adapt their behaviour to each other (Krabben et al. 2019) and act *from* the confluence of the moment-to-moment interaction. This account builds on action theories that argue for the integration of cognitive and automatic processes. For example, Christensen and colleagues (2016) propose that cognitive functions can *directly* influence 'automated' motor execution and hence facilitate skilled performance under pressure. Although muay thai is an extreme example, the case suggests that even in the grim and dangerous setting of a bout, fighters can in fact act mindfully, learn from each other over time and enforce feedback-driven control over well-learned actions.

Muay thai fighting involves a curious balance between collaboration and antagonism, and between competition and strategic adjustment. In the ring, fighters are committed to rules and cultural etiquettes and hence agree on shared ways to compare their skillsets. The extreme yet controlled context allows them to exert antagonistic forces on each other, some apparently cooperative and others obviously belligerent. Such social configurations operate in many sports, but become dramatically palpable and painfully demanding in fleshy, sensuous and violent contact sports such as muay thai.

In the following case study and analysis, I draw on long-term ethnographic fieldwork at the PTJ muay thai gym (short for *Photajaroen*, the family name of the gym owner Andy Parnam's wife), a family run gym in Sydney, Australia. I joined the gym in March 2019 with the aim of collecting data for my PhD. While I draw on my own experiences at the gym, and my own background of practising and competing in different martial arts and combat sports over the last sixteen years, I limit the bulk of my analysis to data from observations and a post-event dialogue with Sing – an expert veteran fighter and coach at the gym.

In June 2019, I had the opportunity to follow Sing as he prepared for a big promotion prize fight, and to discuss the fight with him afterwards. Expert skills in their full-blown ecological settings, where performance matters and the stakes are high, are not easy to capture and comprehend (McIlwain and Sutton 2015). My primary method to tap his thinking on the extremely fast and transient moments of the fight was through an interview dialogue using stimulated recall elicited by video (Lyle 2003); replaying the fight and talking about what happened. My approach is not as narrow as some skill researchers call for in inviting experts to share their thoughts. For example, Eccles (2012) requires valid accounts to comprise self-reports of 'immediate' thoughts ('concurrent reporting') about the target matter. He excludes any interpretation or self-reflection from the athlete. However, as McIlwain and Sutton (2015) point out, additional kinds of knowledge and personal insights other than reported 'concurrent' thoughts are relevant for generating optimal play and responses. As skill researchers have shown through both their own embodied involvement in practice and their engagement with other skilled practitioners, what athletes see and how they battle challenges under different forms of performance pressure is shaped by personal history, context, meaning and knowledge (Bicknell 2010; He and Ravn 2017). In the following, I briefly introduce the field site and practice before presenting the case study and analysis.

What do fighters do?

At the PTJ muay thai gym practitioners learn bodily techniques to interact, play and compete with one another – and, if their skill trajectory suffices, to enter the ring. While many novices begin with the aim of exercising and learning new skills, those who stay quickly get engrossed in the social and skill-driven hierarchical forces of the gym that drive practitioners to want to excel in using their acquired skills in combat.

Before practitioners can progress in their practice there are several perceptual, physical, emotional and motor capabilities that they need to train, manage and consolidate first. For example, to throw a punch or a kick in a face-to-face interaction the attacker needs to be able to recognize an opening or a vulnerability in their opponent, choose an appropriate technique and time their attack at the right moment. Their opponent might be moving in to strike at the same time or trying to thwart the attack by blocking the movement. Because of this dyadic coregulated process, both practitioners regularly have to adjust and change what they are doing, in the middle of a move.

Practitioners need to develop both the perceptual abilities necessary to pick up the relevant information from their adversary's body position and movements so as to anticipate their next step, and the motor capacity to block, dodge, stymy or counter back. It takes expert skill to not only respond astutely to a dangerous, fast and hard incoming blow, but also a physical resilience to pain, alongside well-rehearsed emotional management so as not to wince or reveal nerves in a full-blown confrontation in the ring.

Most novices struggle with perceiving the full span of possible incoming attacks. For instance, one guy I sparred with on occasion at the gym would be so focussed on my legs (and planning his own next move) that I would gently tap him on the head with my glove to direct him to look more upwards. He had not yet built a solid set of defensive perceptual capacities to assess the whole body of his adversary, a skill that can take years to develop. Most novices are typically overly reactive towards possible attacks and become easy victims for deceptive moves. Hence, most skilled practitioners not only build immediate physical reactivity, they also develop precision and astuteness in penetrating deception and distractor moves.

Muay thai encompasses an abundance of heterogeneously categorized styles of fighting. I refer to two dominant fighting methods rehearsed within the muay thai communities I engaged with: 'forward' fighting, also labelled 'pressure' or 'aggressive' style fighting, and 'technical' or 'defensive' style fighting.

The dominant style taught at the PTJ gym is predominantly centred on technical style fighting. Technical fighting is typically thought of as a refined and smarter fighting art compared to the aggressive style. By using a defence tactic, fighters adopt an anticipatory position towards their opponent and aim to evade and 'blend' with the attackers' moves by mimicking their rhythm and tactics. In contrast, another gym at which I also did long-term fieldwork was led by a veteran fighter known for his relentlessness and pressure fighting. Fighting modes that embody aggressive-forward fighting are typically specialized in close-range fighting and adding pressure by tirelessly 'stalking' their opponents. By relying on fitness and endurance in receiving and absorbing many blows, they are typically considered to have less technical finesse compared to technical fighters.

Styles and ways of thinking in the ring are governed by cultural traditions and the unique histories of the fighters and coaches leading the gym. Each fighter comes to embody such cumulative knowledge while, in concert, finding their own ways to succeed, and strategies to draw on, against different opponents.

In dialogue with Sing, we homed in on the key moment in the fight that I described in the introduction. In analysing this discussion, I develop an account of three component stages in low-commitment strategizing in expert fighting. The analysis portrays Sing's perspective in particular, because of the access and extensive fieldwork I undertook with the PTJ gym and Sing. It may be that the adversary engaged in similar processes, though that is beyond the scope of this analysis. The case study suggests that he may not have done so with equal efficiency.

Sing's trap

Sing and I met two weeks after his YOKKAO 40 fight. I brought the recorded fight on my laptop to help elicit his thoughts and direct the conversation towards key moments. Before the fight, we had discussed his expectations of the bout and of the opponent. He and the head coach, Andy, advised that they expected Lloyd to try to 'snatch' the early rounds so Sing would tire later in the fight. They knew that Lloyd was a pressure fighter and, as Andy had noted, that Westerners had a tendency to 'lope' when confronting Thais to 'exploit the physical element'. Their goal was to stay 'outside', stymy any in-fight confrontation, circle around and push kick (*teep*) to control the distance.

This time we met alone to discuss the fight. As we sat down to talk, Sing directed my attention to subtle features that had driven him to execute the uppercut elbow in the second round. He directed my gaze to Lloyd's reactions to the push kicks which Sing had drawn persistently throughout the first round, and leading into the second, enforcing his strategy (Sing executed more than twenty-one push kicks within this timeframe). 'I always push kick, and he is starting to drop his hands to parry my kick like this or like this,' he said. While explaining, Sing mimicked how Lloyd had started dropping his hands to the middle to try to grab and control Sing's thrusting kicks. This was not an action that Lloyd had used in response to Sing's *teep* in their early interactions. Sing continued: ' … he leans forward to not get unbalanced from my kick; he doesn't want to fall back.' Sing showed me how the body placement should be and the effect of the kick if Lloyd had been leaning back in the moment of the kick. Sing added, ' … when someone push kicks you, you have to have the balance: one balance, two you have to stay strong, because if you're too far away, that's a big mistake. He has to show that he is strong and that he can take it.' Finally, at the moment of executing the elbow strike he showed me that he had lifted his knee slightly, faking a push kick. The fake was so subtle and swiftly executed that from a third-person view, at a first glance, I had not noticed it at all. Lloyd in response had leaned into him, bracing himself in expectation of an incoming kick, giving way for the briefest of moments for Sing to land the elbow on his forehead. Sing added that he had planned to execute the feint during the round. It was not an immediate reaction to his opponent's movements in that exact moment: 'That, I thought ahead [...] It's not a reaction. A reaction is just a basic thing: block, kick, block, knee, lean back; the basic thing. But a trick like that, you need to think ahead.'

Sing had not thrown mindless, uncontrolled techniques at Lloyd. Through his technical fight strategy, he had diagnosed a pattern in Lloyd's responses: he probed this strategy several times in the first round and into the second, causing Lloyd to drop his hands and lean into his kick. Finally, he tricked Lloyd to prepare for the push kick and executed the elbow strike when it was too late for him to adjust. Through expert strategic thinking, Sing had managed to lure Lloyd to fall into his trap.

Staging emergent strategizing in the ring

In this section, I demonstrate how the successful elbow strike was realized. I do this by exploring different stages of a dynamic decision-making process; stages I term *diagnosing, probing* and *choosing*. These processes each drive the uncertain, unsettled and unstable development of low-commitment strategies that facilitate skilled performance under pressure.

Diagnosing

In order to develop tactics in an open task environment, an expert must determine what actions are possible and which actions are most likely to be successful given existing constraints. In muay thai, fighters need first to identify information that can direct them towards different opportunities in the interactional continuum.

In this case study, Sing picked up a response pattern in Lloyd's behaviour and noticed that Lloyd was leaning into Sing's kick and trying to push through his defence. Sing identified that Lloyd was committed to his strategy, which opened up a range of opportunities. This identification – which I term *diagnosing* – was not coincidental but emerged from his expert attention and embodied knowledge: *he knew what to look for*.

As ethnographers studying specialized embodied practices have noted, *looking*, in many domains, is a skill that has to be learned in order to pick up relevant information from the environment (Downey 2005; Grasseni 2004). Over time, performers learn to attend to more useful information that aligns with a given goal. This importantly is not the passive reception of information, but 'an active seeking out of important stimuli' (Downey 2005, 34; Gibson 1979). Considering our ability to look a cultural, social and history-dependent skill helps us consider a performer's perceptual skills as part of a motivated, timely, dense and experience-dependent act.

As I hinted earlier, in muay thai interactions, practitioners typically look for trends in the opponents' movements, which is a way for fighters to deal with uncertainty and set-up constraints. In this scenario, both Sing and Lloyd were performing stylistic behaviour; technical-defensive fighting and aggressive-forward fighting. Sing was thus aware of Lloyd's motivation to add constant pressure in order to succeed with his strategy (tiring out Sing). Sing also knew that showing an effect from his kicks by physically getting pushed back would be unfavourable for Lloyd on the judges' scorecard. Moreover, Sing identified that Lloyd had started to lean into his kick and lower his guard to respond to his push kick. This intricate domain- and context-specific background knowledge, I suggest, helped Sing look for and diagnose salient opportunities during the fight.

Looking and identifying opportune moments changes throughout the fight as conditions change, and on a longer timescale may be conditioned by a fighter's changing circumstances. In our dialogue, Sing pointed to the fact that his transition from fighting full-time in Thailand to moving overseas to Sydney and coaching full-time instead had driven him to change his style from forward fighting to technical style fighting. He had decided to make this change due to his limited training time and thus limited stamina to endorse the pressure needed to retain an aggressive strategy.

Building on cognitive ecological theory, we may understand Sing's diagnostic skills as hosted by distributed knowledge of the unstable situation unfolding. His knowledge of the series of micro-events leading up to the fake consisted of a deep-seated understanding of his opponent's fighting style, movement trends and motivation in relation to scoring. His specialized gaze was also directed by his embodied knowledge of his own skillsets and changing life situation. While a fighter may be able to acutely perceive various different openings, what is diagnosed as salient are options (patterns of stimuli) that are relevant in the moment and available as possible items to act on.

Probing

Opportunities in dynamic antagonistic interactions such as martial arts are often described as transient: 'they evolve and devolve again' (Kimmel and Rogler 2018, 3). To cope with such fleeting moments, fighters will actively test and nudge discovered opportunities to create, alter and sustain them instead of immediately acting on them. In the second stage, I map a set of techniques common to muay thai that I call *probing* to address the feedback-driven aspects of low-commitment strategizing.

Probing is a technique that reveals information by actively provoking and testing different responses. In fight interactions, probing is done by physical stimulation. From a naïve perspective it may appear as if a fighter can strike or be struck at any moment. But in a high-level and well-matched fight interaction, fighters need to move through action 'phases' to break the other's defence. Muay thai fighters call the early phases before implementing a technique a 'setup' which consists of preparatory moves that serve to distract, conceal or mislead the opponent from guarding or preparing themselves for an upcoming action. Preparatory moves are rehearsed on different timescales from immediate tricks, drawing on common habits and training histories, to longer scales over one to several interactions that purposely work to probe for and seek out certain responses for manipulation.

As mentioned, Sing had diagnosed an opportunity in Lloyd's response pattern and had continued to push kick him several times, until the moment of executing the elbow strike. While this was a defence strategy to 'shut down' and control Lloyd's forward pressure, I suggest that it also served as a mechanism for exploring and generating options, a process that works iteratively and in concert with diagnosing. Probing helped direct Sing to important features in the interaction and actively create opportunities based on the generated response pattern. An evolving action plan could start taking shape as the fighters' interactions actively created constraints and opportunities for each other.

Interestingly, it is the mutuality of this encounter that permits Sing's unfolding plan. Lloyd, who was also adjusting to Sing's defence by attempting to resist and catch Sing's kicks, was responding with greater commitment over time; moving into the second round Lloyd was increasing pressure to dismantle Sing's defence and was pushing in more aggressively. He thereby strengthened the response pattern and the reliability of the response. In fact, Lloyd was adapting better to Sing over time (getting pushed back less), which paradoxically contributed to him setting himself up more.

This reciprocally constructed moment may be further characterized as involving processes of entrainment. According to Clayton, 'entrainment is the process by which independent rhythmical systems interact with each other' (2012, 49, cited in Geeves, McIlwain, and Sutton 2014, 1). This process can lead to a stabilization phase: a temporal coordination between the interacting parts even when coordination is unintentional or under unfavourable conditions (Geeves et al. 2014). For example, Richardson and colleagues (2007) found that people sitting next to each other in rocking chairs unintentionally fell into synchrony even when the natural frequency of the two rocking chairs differed. Clayton (2012) emphasizes that entrainment can be asymmetrical; an agent with greater power or dominance can exert a disproportionate amount of influence on the interaction, as in a leader–follower scenario.

Processes of entrainment reverberate into social interactions where a self-organized pattern may be formed, such as in the case study. Yet, in the fight scenario, these processes are not a result of either cooperative or unintentional forces, but rather a volatile, 'collaborative antagonistic' configuration.

From this perspective, we might understand probing as an active tool used to steer an interaction towards a moment of criticality; which in this case was the moment of the feint. Probing is the development of a portion of a plan by actively directing and nudging action choices. Sing actively provoked the situation and revealed information about Lloyd's responses and behavioural tendencies. He probed diagnosed opportunities with a flexible, responsive logic.

Choosing

Action choices in close-range interactive duelling sports such as muay thai are dependent on and constrained by what the other is doing at the same moment.

The final implementation of an evolving strategy is therefore precarious and has to be negotiated. In the last stage, I suggest that there may have been a tipping point where Sing identified that there was a reliable negotiated effect that led him to execute the elbow technique: he chose the right moment to enter into a dangerous close-range attack.

In muay thai an uppercut elbow is thrown by bringing the elbow from down to up in a slicing movement through the middle in a vertical plane. It hence necessitates confronting the opponent at close proximity – the target literally at an elbow's distance. An elbow strike is therefore dangerous to implement, especially against an aggressive forward fighter who is already seeking to dominate in the in-fight.

As Sing emphasized in our dialogue, the implementation of the elbow strike was not a 'reaction'. It was something he had perceived as a *possibility* by diagnosing and probing – *after* recognizing how Lloyd was responding to his push kick. Sing's action choice was not an isolated event, but generated from the prior stages and enforced through a co-generated anticipation process which he used to trick Lloyd to commit to the wrong move and break through his defence.

In sports psychology, studies on skilled performance suggest that experts, compared to novices, have sophisticated prediction skills that allow them to control not only their own actions, but also those of their opponent and the greater contest (MacMahon and McPherson 2009). Action choices are not always simply reactive, but at a high level of expertise, as this case shows, are informed by a rich knowledge base, and by flexible and fast evaluation of coregulated embodied interaction. Commonly within action theory, improvised actions are regarded as singularly isolated events, and generally overlooked as concatenated, structured entities (Preston 2013). *Choosing*, as this case depicts, directly connects with and follows from prior action events and is therefore an appropriate term for the third and final identified stage of a dynamic decision-making process in enhancing the intelligence of the action choice. As Sing alludes to himself, a moment like the one illustrated above is the opposite of an automatic response. He created that moment himself.

Probing in skilled (antagonistic) performance

The precariousness of skilled performance in variable and challenging conditions has led some expertise researchers to explain skilled decision-making as organism-environment attunement. Following trends in ecological psychology led by Gibson (1979), Silva and colleagues (2013, 767) suggest that actions are shaped and constrained solely by a trained capacity to skilfully attune to regularities in the environment. They write that 'the environment is perceived directly in terms of what an organism can do with and in the environment [...] it is not dependent on a perceiver's expectations, nor mental representations linked to specific performance solutions stored in memory.' It is challenging for such an account, which leaves no room for cognition or thinking in action regulation, fully to explain the nature of skill.[2] Can Sing's smooth, emergent and enacted strategy be explained solely by attunement to current ecological constraints of action opportunities? Or may we regard his actions as intelligently constructed, albeit in close exchange with the unfolding situation? Although these questions will not be solved here, my analysis leans towards the latter. As I have shown, fighters, at high levels of skill, can sometimes actively direct unstable situations by mapping opportunities through *diagnosing*, directing and nudging action choices by *probing*, and *choosing* reliable moments to act.

Probing is an active method and process that is highly sensitive to the triadic agent–agent–environment relation, yet involves employing and generating feedback towards self-enabling ends. It thereby undergirds the idea that expert athletes can think strategically in action and follow a low-commitment 'plan', one that is not preconceived or rigidly laid out in advance.

After Sing implemented the elbow strike it dramatically changed the dynamic of the fight. The cut on Lloyd's forehead inflicted by Sing opened up a new profile for both of the fighters: in the rest of the round, Sing was noticeably trying to open up the wound more and end the fight, while Lloyd, who was still trying to resist and push forward, was required to protect the gushing wound so as not to be stopped by technical knockout. Sing had not only managed to implement a dangerous technique; his trap created a new terrain for both fighters and a new set of challenges in the ongoing interactional contingency of the fight.

Conclusion

Sing's trap demonstrates how even under extreme conditions, skilled athletes can act mindfully under pressure, and construct and follow fragile, co-generated threads to successful moments. Although muay thai requires trained resistance to pain, emotional resilience, technical finesse and physical endurance, fighters are also aware of their abilities and need to steer and regulate a volatile interaction in smart and cunning ways. Such capacities for awareness and decision-making depend profoundly on community norms, and are entrenched in long histories of practice.

This chapter has specifically focussed on high-level cognition in challenging, forceful and fleshy fighting practices. I have mapped three stages of low-commitment strategizing in the ring: *diagnosing*, *probing* and *choosing*. I have discussed the 'active' way a skilled fighter may problem-solve in the ring, and have briefly discussed implications for theories of organism-environment attunement.

Case studies drawn from close cognitive ethnographic examinations of antagonistic fighting practices can help us access richly embodied skills as they are expressed in real complex ecologies of practice. Observations 'from' and not purely 'about' a practice, in this case, demonstrate the cognitive demands on practitioners and the fluent and environmentally iterative ways that they confront and control such challenges. Such striking processes remain hidden when we do not engage with and attempt to understand such accounts from the inside.[3]

Notes

1 The complete fight is available on YouTube: https://www.youtube.com/watch?v=z7-XR47QvEA&t=958s (YOKKAO 2019). The elbow strike occurs at 13:07.
2 For further discussion on this issue, please see McIlwain and Sutton (2015) and Christensen and Bicknell (2019).
3 I thank Ian Maxwell, Kristina Brümmer, John Sutton, Kath Bicknell, Greg Downey and Aitor Miguel Blanco for their constructive feedback. I also thank Singpayak and Andrew Parnam for their time and insights from which the ideas in this chapter could be brought to life.

References

Bicknell, K. (2010), '"Feeling Them Ride": Corporeal Exchange in Cross-Country Mountain Bike Racing', *About Performance 10*: 81–91.

Christensen, W. and K. Bicknell (2019), 'Affordances and the Anticipatory Control of Action', In M. Cappuccio (eds), *Handbook of Embodied Cognition and Sport Psychology*, 601–22, Cambridge, MA: MIT Press.

Christensen, W., J. Sutton and D. J. F. McIlwain (2016), 'Cognition in Skilled Action: Meshed Control and the Varieties of Skill Experience', *Mind & Language 31* (1): 37–66.

Clayton, M. (2012), 'What is Entrainment? Definition and Applications in Musical Research', *Empirical Musicology Review 7* (1–2): 49–56.

Downey, G. (2005), *Learning Capoeira: Lessons in Cunning from an Afro-Brazilian art*, New York: Oxford University Press.

Eccles, D. W. (2012), 'Verbal Reports of Cognitive Processes', In G. Tenenbaum, R. Eklund, and A. Kamata (eds), *Measurement in Sport and Exercise Psychology*, 103–17, Champaign, IL: Human Kinetics.

Engel, A. K. (2010), 'Directive Minds: How Dynamics Shapes Cognition', In *Enaction: Towards a New Paradigm for Cognitive Science*, 219–43, Cambridge, MA: MIT Press.

Geeves, A., D. J. F. McIlwain and J. Sutton (2014), 'The Performative Pleasure of Imprecision: A Diachronic Study of Entrainment in Music Performance', *Frontiers in Human Neuroscience 8*: 863.

Gibson, J. J. (1979), *The Ecological Approach to Visual Perception*. Boston: Houghton Mifflin.

Grasseni, C. (2004). 'Skilled Vision. An Apprenticeship in Breeding Aesthetics', *Social Anthropology 12* (1): 41–55.

He, J., and S. Ravn (2017), 'Sharing the Dance – on the Reciprocity of Movement in the Case of Elite Sports Dancers', *Phenomenology and the Cognitive Sciences 17* (1): 99–116.

Kimmel, M., and C. R. Rogler (2018), 'Affordances in Interaction: The Case of Aikido', *Ecological Psychology 30* (3): 195–223.

Kimmel, M., and C. R. Rogler (2019), 'The Anatomy of Antagonistic Coregulation: Emergent Coordination, Path Dependency, and the Interplay of Biomechanic Parameters in Aikido', *Human Movement Science 63*: 231–53.

Krabben, K., D. Orth and J. Van Der Kamp (2019), 'Combat as an Interpersonal Synergy: An Ecological Dynamics Approach to Combat Sports', *Sports Medicine 49* (12): 1825–36.

Lyle, J. (2003), 'Stimulated Recall: A Report on Its Use in Naturalistic Research', *British Educational Research Journal 29* (6): 861–78.

MacMahon, C., and S. L. McPherson (2009), 'Knowledge Base as a Mechanism for Perceptual-Cognitive Tasks: Skill is in the Details', *International Journal of Sport Psychology* 40 (4): 565–79.

McIlwain, D., and J. Sutton (2015), 'Methods for Measuring Breath and Depth of Knowledge', In J. Baker and D. Farrow (eds), *The Routledge Handbook of Sports Expertise*, 221–31, New York: Routledge.

Preston, B. (2013), *A Philosophy of Material Culture: Action, Function and Mind*. New York: Routledge.

Richardson, M. J., K. L. Marsh, R. W. Isenhower, J. R. L. Goodman and R. C. Schmidt (2007), 'Rocking Together: Dynamics of Intentional and Unintentional Interpersonal Coordination', *Human Movement Science* 26 (6): 867–91.

Silva, P., J. Garganta, D. Araújo, K. Davids and P. Aguiar (2013), 'Shared Knowledge or Shared Affordances? Insights from an Ecological Dynamics Approach to Team Coordination in Sports', *Sports Medicine* 43 (9): 765–72.

Van Dijk, L. and E. Rietveld (2021), 'Situated Anticipation', *Synthese* 198 (1): 349–71.

YOKKAO. (2019), 'YOKKAO 40: Singayak PTJ MuayThai Vs Lloyd Dean I Muay Thai -65kg I Full Fight', 6 February 2021, *YouTube*, Available online: https://www.youtube.com/watch?v=z7-XR47QvEA&t=958s.

10

Intercorporeal synergy practices – perspectives from expert interaction

Michael Kimmel and
Stefan Schneider

A roadmap to qualitative research

Synergy in Greek means 'working together'. We often use the notion colloquially, but in the movement sciences it carries a technical meaning about coordination patterns: It implies a study of how elements in a movement or interaction system are temporarily coordinated for specific tasks (Latash 2008; Turvey 2007). Synergy scholars generally assume that behaviour emerges from a (non-hardwired and variable) interplay of elements, which are dynamically assembled to fit the task and the context. This implies an *interdependent* control of parts within wholes. Since the synergy concept is essentially 'scale-free', these parts can be neurons, muscle fibres, limbs, but also several individuals, depending on the research focus.

This contribution proposes a qualitative inquiry into synergies that are created in interaction, i.e. where two individuals provide dynamic ecologies for each other. At this level we investigate how multiple bodies can coordinate 'as if they were one' as they accomplish particular actions together. The phenomenon is familiar enough. We routinely create interpersonal synergies as we engage in conversations, carry objects together, smoothly move through busy traffic, make love or music together or enjoy ballroom dancing and team sports. Efficient ensemble behaviour and even interpersonally 'distributed' forms of creativity pervade many of our experiences.

We here present a micro-analytic approach that reconstructs the structural basis and assembly dynamics of interpersonal synergies. Readers can expect a peek into the intricacies of synergy regulation, viewed through the eyes of interaction experts. This is based on short vignettes, which make the subjective 'action logic' of each participant accessible and, on this basis, explicate how synergies come into being from the recursive interplay of action and response between them. For our analysis we have sampled highly skilled forms of interaction operating 'at close quarters' in which physicality extends across the skin boundary and builds on continuous rapport and sensorial interpenetration. In approaching our topic, we will contrast different collective-level 'synergy logics' that underlie acrobatics, dance and martial arts, respectively.

Background and objectives

When we apply the notion of synergy to these *intercorporeal* interaction forms this means investigating how elements from multiple bodies connect and synchronize for a biomechanic ensemble function.

Analytic definitions

Araújo and Davids define interpersonal synergies as a 'collective property of a task-specific organization of individuals, such that the degrees of freedom of each [...] are coupled, enabling [them] to coregulate each other' (Araújo and Davids 2016). Accordingly, two or more individuals

- create an *ensemble functionality* which exceeds mere additive or externally coordinated behaviour, say, as when a group of people run for shelter in a thunderstorm.
- restrict their individual degrees of freedom under the demands of a task-constrained global organization, manifest in coordinated geometries, rhythms, distances, etc.[1]
- make interdependent adjustments to establish and maintain this organization, e.g. through compensation of partner glitches in collaborative interactions or by counteracting hindrances and disruptions in antagonistic ones.

In brief, elements from different individuals enter into an interdependent organization, establishing a macro-system with some collective functionality. This can imply collective *purpose*, but not necessarily. In non-collaborative interactions, as we shall see, synergies arise despite substantial efforts of another person to prevent them.

Synergy practices

It is consensus that interpersonal synergy research should 'identify the process in which system components vary to stabilize task specific performance variables' (Passos et al. 2018). Current research is dominated by behavioural experiments, e.g. on synchronization (Harrison and Richardson 2009; Riley et al. 2011; Schmidt et al. 2011; Schmidt et al. 1990) and simplified, but ecologically more valid applications to team sports like rugby or soccer where metrics have been developed to investigate changes in collective dynamics, work-sharing patterns, preferred communication channels (Duarte 2012), degrees of synergy (Passos et al. 2018) or coupling modes (Bourbousson et al. 2010). These approaches focus on somewhat aggregate data patterns.

We presently propose a complementary qualitative perspective. We emphasize that, since synergies arise from the interplay of *parts* relative to task *wholes*, our analysis should co-equally address these levels. Accordingly, we take interest in macro-parameters that ensure collective task functionality (i.e. performance variables), take stock of contributing micro-elements and describe the patterns of interplay, i.e. linkages and sharing relations (Araújo and Davids 2016; Riley et al. 2011), which connect these analytic levels. The central questions thus are which elements participate, how they enter into couplings and how they create a collective function together.

Our specific aim is to investigate *synergy building practices* at the level of short interactions. Our reconstruction is based on interviews with expert couples and what they tell us about their individual micro-scale perceptions and actions as they skilfully engage with each other. The analysis then focuses on how one body connects with the other, how collective structures are created, how feedback is used to adapt them and how the whole is kept constrained enough to work. The micro-analytic methods that we employ (introduced in Kimmel 2021; Kimmel and Hristova, 2021) involve a dialogue in which 'thin slices' of video footage are jointly reviewed ('stimulated recall'). Complementarily, phenomenological interviewing techniques are applied, helping the interviewees to arrest attention on one moment at a time, to verbalize tacit aspects and to progressively probe into details. The comprehensiveness and high zoom factor of this approach allows us to track how the experts build, develop and manipulate synergies as the interaction unfolds.

Our present contribution explores two complementary aspects, *synergy structure* and *synergy dynamics*. The structural focus tries to capture how coupled elements from two contributing bodies create physically effective macro-functionalities, while the dynamic focus explores the assembly process and evolution of these synergistic macro-functionalities, which are mediated by embodied communication.

Intercorporeal structures

Dance, martial arts and acrobatics operate at close quarters. Whereas soccer and rugby teams or everyday activities, such as walking past another person in a narrow corridor (Di Paolo and De Jaegher 2007), have a dominant proportion of informationally mediated synergies, the practices we present here additionally exhibit more lasting structures of 'mutual incorporation' across the skin barrier (Froese and Fuchs 2012). This section aims to discuss connective inter-body structures in their physicality and organizational logic.

Collective physics

Intercorporeal structures, as defined here, display a shared functional physiology. This 'collective physics' can, e.g. include any of the following architectural aspects:

- weight sharing
- counterbalance
- 'stacked' skeletal alignment
- connective muscle arcs
- kinaesthetic vector configurations, levers
- stability arrangements of skeletal struts + elastic fascia (*tensegrity*)

A collective structure, as we shall later see, can include several of these principles at the same time and combine them in sophisticated ways. Understanding the collective architectural functionalities has relevance for teaching, e.g. to create didactic imagery, but also for researchers to model how complex collective structures arise. This can shed light on how synergies come to create macro-functions for collective action, e.g. what tango dancers metaphorically call 'creating the beast of four legs', and which connections across the skin boundary contribute here. This analysis may ultimately suggest domain-specific principles of *structural intelligence*.

The logic of joint building

For illustration, consider a figure from acroyoga, a species of partner acrobatics. In the example (Figure 8) one architectural principle is superimposed once a more basic one has been established between the partners.

The initial principle is termed 'stacking' and it is complete as soon as the person on top, the 'flyer' (F), assumes a shoulder stand on the arms of the 'base' (B). This principle operates by positioning F's weight above B's aligned arms, which directs this weight into the ground via his shoulder girdle. This configuration produces a fairly self-stabilizing position as long as F holds her body form. A moment later they begin to introduce instability again; they complexify the configuration by adding another architectural feature: B brings his

 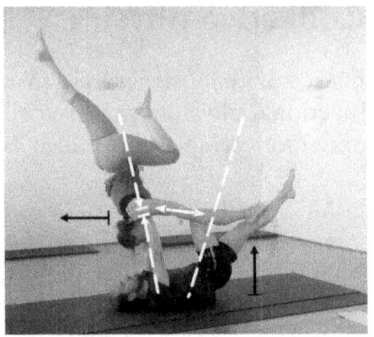

FIGURE 8 *Acroyoga counterbalance (a) aligned 'stacking'; (b) opening a triangle with a pull connection. Photos by M. Kimmel.*

hands a bit over his head so that F overshoots the (self-supporting) line of alignment. This instability is compensated for by F's hand grip on B's leg, which introduces a pull connection to his lower body. As soon as the flyer goes into the outlier position, her weight pulls his hips and legs up on the other side as a counterweight, with his upper back acting as fulcrum. Only the combination of off-balance stacking, the pull and counterweight together ensure overall architectural stability again. The reported macro-imagination is that of counter-balanced scales, which the practitioners describe as an *inverted A* (cut off at its base) that progressively opens the sides.

Our analysis thus takes stock of elements (e.g. body-part actions) to understand their local coalitions (e.g. alignment lines of B's arms and F's body line), the mutual constraint or complementariness between sub-synergies (e.g. allowable range of off-balance without losing support stability) and the composite subfunctions (e.g. alignment, supports, struts and connecting vectors of elastic pull) that define the emergent constellation.

Sharing qualities and degrees of coupling

A diversity of sharing patterns can underlie collective structures. This specific 'division of labour' between agents can – often depending on the domain – include distinctions such as parallel vs. alternating actions (Bell and Kozlowski 2002), sequential vs. parallel initiation, symmetric vs. complementary movements or mere element aggregation vs. deep fusion (Araújo and Davids 2016). There can be *coalitions*, i.e. body segments temporarily 'frozen' to move together (Turvey et al. 1978), various directions from which a linkage is built (Kimmel et al. 2015) and different co-dependency relations such as mutual facilitation, constraint or competition between elements (Kimmel 2017).

Some synergies may involve very strict coupling requirements. Others remain selective and allow subordinate autonomy or even partial decoupling. A tightly fused synergy created the above acrobatic lift, while other synergies allow for some independence. We know from solo skills that sub-movements can be co-specifying, yet also biomechanically isolating, e.g. when a dancer's legs perform multiple weight shifts while the torso stays immobile. Similarly, collective coordination can allow independence within bounds: For example, in dancing Argentine tango, followers can autonomously 'fill in' adornments with their free leg as long as they respect the leader's signal for shifting the leg. Within this short time window the adornment is a *parallel* individual action. The observer sees an *aesthetic* synergy of two complex actions performed at the same time, yet with considerable biomechanic autonomy.

More generally, collective structures frequently admit of *subordinate autonomy*, e.g. when a given performance parameter allows multiple ways of executing it. So the investigation needs to move to (a) degrees of general constraint and (b) variation ranges in interplay patterns.

Embedded synergies and upward regulation

In the interpersonal context, particular individual movements function as *embedded regulators* within larger inter-body synergies. That is, some components are used for local purposes by individuals, while others are strategically employed to 'upward regulate' the inter-body coupling. This means that individuals selectively feed energy or information into the collective physics. To exemplify selective extension, let's take the martial art aikido (Kimmel and Rogler 2019): Defenders need to simultaneously maintain their own upright axis in their core body and use their movement in space and body periphery to progressively 'get into an attacker's axis' by applying a lever that blocks the wrist, then the elbow and finally, as the shoulder joint is impacted, allows manipulating the attacker's balance. When one uses internal synergies to regulate a collective whole, a good compromise needs to be found that works at individual and collective levels.

Habits as synergy substrate

The broadest kind of embedding structure for synergies are domain-specific *body habits* that expert practitioners readily supply the very moment they begin. Habits furnish a baseline for all manner of more demanding interpersonal synergies. They are prerequisites that often remain invisible unless one contrasts this with the performance of novices who lack the basic self-structuration. To illustrate, tai chi practitioners habitually stay upright, grounded and cultivate a connection of parts, where large lower-body muscles generate force, hips and waist give it horizontal direction, while upper extremities merely 'express' it. This organization prepares for partner exercises, in which one pushes from the ground and through one's whole body structure, or channels force received from the partner downwards through the same habitual organization.

Individual habits thus prefigure ways of interpersonal connecting. They reduce degrees of freedom in establishing a collective system and to simplify its control once it is in place. Structural pre-organization influences whether an impetus from the other person gets channelled into the body, gets 'smothered', deflected, elastically rebounds or adds to a movement. Pre-organization also makes for collective action readiness. It prepares the ground for complex interpersonal synergies to kick in by pre-activating basic structures that can be quickly added to and extended across the body boundary.

Likewise, a good basic relational organization with others (i.e. interaction habits) can supply foundations for a wide range of more complex synergies. A tango embrace exemplifies this, a chest-to-chest connection with a small amount of weight sharing. It supplies the basic 'rapport synergy' which affords quick recognition of signals, swift impulse transmission, as well as sensing and being sensed from the torso interface down to the legs. All further coordination tasks of the dance – especially well-timed footwork – benefit from the embrace's quality.

Synergy dynamics

A complementary task is to model the time course of how intercorporeal synergies evolve at short timescales. Here, coupling loops between agents (Araújo and Davids 2011; Passos et al. 2016) move into view whereby two persons build, develop and manipulate collective synergy structures through a process of 'give-and-take'.

Micro-assembly of synergies

The proposed *process audit* poses two complementary challenges: (a) to micro-dynamically describe the synergy evolution in its own terms, and (b) to reconstruct how a moment-to-moment process of embodied communication brings forth the interaction trajectory 'carrying' this synergy. We start by putting micro-actions on a timeline, before working our way up to interactive causalities as one micro-action triggers the next. This involves modelling two important aspects of the process:

- Which actions are created to *initiate, add to* or *modulate* interaction processes? How does an action impact the other person or the collective milieu? How are existing collective physics manipulated?
- Which perceptual *affordances* (Gibson 1979) do individuals create for each other (whether deliberately or not),[2] such as start or continuation triggers, invitations and assistance requests, requests for adaptation, stop/alarm signals or lures?

Summarily, a process audit should reconstruct how, from the mutual reactivity of micro-strategies, an interaction trajectory emerges through which a collective synergy is co-assembled and developed. This should, in turn, allow us to identify the more general process logic whereby a synergy evolves, which can, e.g. involve a stepwise complexification, as in our acrobatics example, continuous pattern complementation, playful variation or a dynamic of vying, as the following analysis of an antagonistic interaction will exemplify.

The logic of thwarting, luring, deceiving and converting

All 'soft' martial arts show a dynamic of intense synergy negotiation between an attacker and defender. They follow opposed intentions and create physical antagonism, which make for distinctly non-collaborative synergy dynamics (even if, in training exercises, teachers may not always fully exploit their dominance). The winner imposes the desired synergy to 'selfish' ends by counteracting or breaking the opponent's desired synergy.

Our showcase is a scenario of tai chi push-hands ('tui shou'). The basic pattern is a continuous push-and-yield cycle (*'stick, adhere, join, follow – don't resist, don't disconnect'*). Through the contact point of the arms practitioners can detect even minimal changes in the opponent's body organization and centre of gravity, which helps their whole body to 'stick' to the other's movement. A tight coupling allows (a) for rapid self-adjustments in relation to the opponent, and (b) for the immediate detection of affordances offered by the opponent and their rapid conversion into physical leverage.

As long as both participants are perfectly grounded, a cyclic dynamic occurs – both convert the opponent's pushes into counter-pushes. To break this *symmetry* (Richardson et al. 2016), the opponent must be thrown off-balance through subtle manoeuvres, while maintaining one's own balance and keeping movement options intact. Consider the scenario in Figure 9: Feet positions are fixed; weight shifts between the feet and transversal rotations of the torso are used to interact. An interaction sequence with two main phases ensues.

In images 1–4, the attacker (A, on the left) pushes forward by shifting his weight and trying to use both arms arm to get hold of the defender (D, on the right). A aims to connect with D's body centre in order to push her off-balance. D reacts by yielding to the incoming force so as to lure A into overshooting his balance. D continuously moves her body centre and weight *backwards*, with the defending arm moving along. The immediate effect on A's perceived affordances is an impression of 'nothing there to push'. Simultaneously, D rotates her torso to the right side with an outward movement of her defending arm. In image 2, A has lost contact of his hand to her elbow. In images 3 and 4, A's stability begins

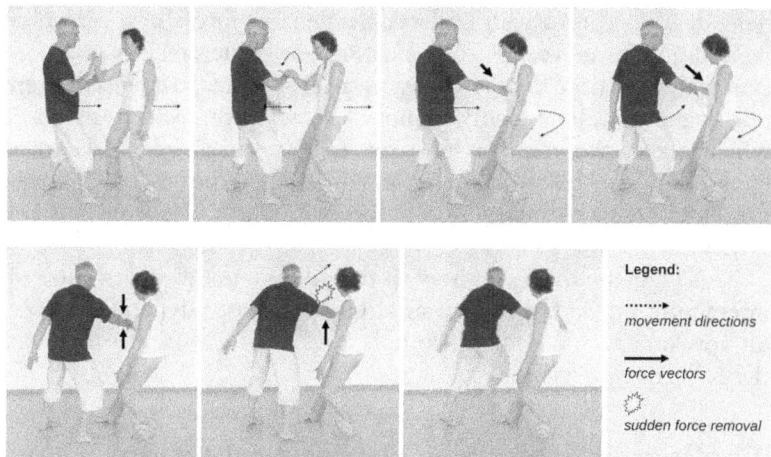

FIGURE 9 *Push-hands: (a) 1–4 'push-yield', (b) 5–7 'retreat-pluck'.* Photos courtesy of Loni Liebermann and Hella Ebel.

to be compromised through his continued pushing. His frontal overshooting reaches a point where D can pull him further ahead/downwards and where her spiralling movement twists his internal geometry sideways, almost ready for tilting. In images 5–7, A struggles to regain the compromised balance: He actively intensifies the push against the 'faux'-stable point to restabilize himself while trying to withdraw his centre of gravity. D now uproots A's balance by turning this action against him. D's hand on his forearm impedes A's withdrawal; she then lifts the downward pressure from his arm. The synergistic arc that supports him disintegrates, an action known as 'plucking'. A suddenly 'leaps'. Thus, D's internal synergies and 'groundedness' are extended into a collective architecture and employed to effortlessly manipulate the opponent by using his own pushing and forward-leaning force against him.

The overall strategic logic is to 'turn around' the attacker's push to realize one's own winning synergy. This works by exploiting forces the opponent supplies, while also inducing actions that will eventually backfire: In terms of subtle perceptual inducement the defender encourages overshooting by presenting the attacker with a lure, a 'false' affordance of structural stabilization against a resisting force. In terms of physical manipulation, the defender (a) progressively deflects push energy into a new direction by adding a force vector to an existing one, which results in a new summary direction, (b) exerts downward resistance to incite counter-pushing and (c) suddenly removes forward-directed resistance, which makes the attacker fall over. The attacker's loss of balance in the final constellation is the outcome of mutual reactivity over time and all prior interaction stages with their respective micro-synergies.

This vignette illustrates how to transduce a micro-description of interaction into a theoretical analysis of synergy dynamics, based on how one person adds to, complexifies or otherwise manipulates the collective physics.

The logic of dynamic complementation

Another pattern of synergy dynamics is that of *ongoing* micro-complementation in a flow of changing, yet perfectly connected co-actions. These synergies, typical of dance or music improvisation, require precisely timed 'co-actualization' of role-specific constituents from both individuals.

Consider how a tango step is executed (Kimmel 2019), where the leader walks forwards and the follower backwards. A precisely micro-coordinated co-assembly pattern underlies this. The follower's free leg makes space for the leader's leg as soon as he projects energy forward through controlled imbalance. The follower's upper body mustn't budge yet. Then, the leader releases his energy. A shared weight system is created, in which the follower 'collects the leader's energy' for a connected movement.

The question is how this continuous complementariness in a joint step works at the control plane. Tango dancers train specific contingency rules for translating what the partner signals through the joint embrace and axis shifts. Leaders combine multiple 'micro-tools' to customize their lead, such as the degree of axis shift, shoulder opening, torso twist or upward lengthening. Each micro-tool triggers a specific tiny response in the follower. The variable mixes of these *primitives* employed by the leader provide for myriad inter-body synergies, e.g. playing with a back-and-forth of the follower's legs or combining a step with elements such as leg invasions or hooks. Each micro-tool modulates the coordinative pattern in specific ways; by combining multiple such micro-tools leaders can create complex effects.

Spontaneous micro-assembly

A related question concerns how creative synergies emerge. We looked at how improvisers converge on unpremeditated synergies through processes of mutual enablement, assistance, complementation or counterproposal. Figure 10 illustrates this through a spontaneously emerging lift in Contact Improvisation dancing (Kimmel 2021; Kimmel and Hristova, 2021): Dancer A circles around dancer B who is placed on all fours, her hands loosely connected to his shoulder. The moment A stops at B's head, an aligned configuration emerges that both perceive as what we would term a 'synergy kernel'. B projects himself upwards into the inner structure of A, who complementarily transfers her weight to his shoulders until she lifts off. How, then, are they able to mutually 'home in' on this interpersonal geometry? B habitually aligns the skeletal lines of arms and legs, so stable force lines running to the ground are already in place, for A to spontaneously 'superimpose' her own aligned structure. The resulting inter-body architecture is largely self-supporting and sends A's weight through B's 'base-structure' into the ground. To convergently assemble a viable synergy, the partners need to respect the emerging collective physics and understand generic functionalities of lifting. The synergy progressively gains directedness through the micro-negotiation between the two dancers (providing 'green lights' for one another, etc.), who jointly complexify an initially less demanding synergy. All this is possible because both dancers recognize synergy potentials, i.e. 'kernels', suitable for augmentation and perform complementary actions that sufficiently respect the macro-structural logic of the emergent task, yet work with it creatively.

To sum up, we can speak of a *softly assembled* pattern that is genuinely distributed across individuals. A collective macro-function is progressively formed between the partners through continuous complementation with emerging directedness,[3] yet also openness. This spontaneous co-assembly process is no easy feat. It presupposes reactivity, instantaneously recognizing what is afforded, and activating a differentiated repertoire of primitives

FIGURE 10 *Contact Improvisation lift. Photos by M. Kimmel.*

'on-demand', selectively mixing and recombining them within the constraints of the emerging task-dynamic logic.

Note that co-assembly processes are frequently subject to a general process logic, which delimits the range of creativity. For example, the aikido aim to 'take an attacker down without injury' obeys an integral logic of progressively narrowing down the opponent's degrees of freedom. Yet, its realization in space, its timing and even the preferred lever technique depend on the interaction dynamic (Kimmel and Rogler 2019): for example, with late initial timing a defender will resort to larger circular movements that give time for 'diluting' the disadvantage. The initial conditions and the mutual reactivity between individuals explain the preferred mix of micro-tools. Note that a process logic that narrows down the range of possibilities can temporarily apply to examples like the above Contact Improvisation lift. The incipient dynamic imposes strict safety and co-functionality constraints until the feet are on the ground again.

Adaptation, variation and contingency

Creating synergies together frequently requires making continuous adjustments to fluctuations and perturbations. Agents use task-protective adjustment strategies to maintain or optimize a synergy, or adjust to compensate for errors (Riley et al. 2011). For example, an aikido lever can meet resistance in the opponent, so either additional abdominal muscles kick in – a force-based strategy – or the practitioner exploits this resistance by switching to the opposed lever (Kimmel and Rogler 2019). Adaptive repair dynamics can thus recruit a further synergy component or actively give the synergy a new direction that opportunistically exploits the given forces.

Likewise, the inherent contextuality and variability of tasks needs to be skilfully handled. Experts know how to adapt at the micro-level to preserve the macro-logic. To exemplify, a Contact Improvisation expert explained how to optimally adapt components relative to known contingencies in doing a backflip over a partner. If the lifted person gets faster the 'base' must scale up muscle tone; the more the lifted person raises her legs and arms – thus increasing tone and momentum – the more micro-balancing of the 'base' is needed; the more the lifted person's head is relaxed the less effort of the legs is needed by the lifter. Our expert informants are typically aware of the multiple situated 'what ifs' of a scenario and know the appropriate strategies of contingency management.

Task transitions

In synergy transitions one holistic pattern devolves and a new one emerges. Some transitions follow the natural desire to move on to the next task, e.g. in improvisational *synergy flows* (Kimmel 2021). Others are motivated by errors or saves. Especially when attempts to repair a dwindling synergy become difficult, rerouting on the fly to a more energy-efficient alternative can seem preferable (Kimmel and Rogler 2018).

Transitional strategies can display numerous dynamic patterns. Many transitions gradually 'ease into' the next synergy; they 'morph' the synergy with some parameter overlap. However, sudden synergy shifts also occur (Torrents et al. 2016). This requires addressing whether sharing patterns between synergy components are extended, reduced, morphed or replaced *en bloc*. Some strategies selectively reparameterize existing components or bring unused ones into play. Others reorganize the dynamic as a whole through new components and sharing patterns.

Conclusion

Our present aim was to discuss intercorporeal synergy practices and provide a number of focalizing concepts. In terms of synergy structures, we took stock of elements that a synergy consists of, described collective physics that are established across bodies (many of them with an 'architectural' character) and tracked how collective physics are skilfully manipulated to reach desired aims. In terms of synergy dynamics, we aspired to a high zoom factor in terms of 'who does what when' when assembling, developing or adapting a synergy. We also showed how synergies need embodied communication to emerge.

On the way, we contrasted different *synergy logics* that vary with the domain-specific aims. Instructive contrasts emerged here: First, some synergies are routines with familiar sharing patterns, while others are spontaneous. With largely 'pre-packaged' synergies only timing and micro-tuning are subject to interactive negotiation. Secondly, collaborative and antagonistic domains differ. When agents collaborate, the synergy exhibits mutual purpose; accordingly, agents complement and compensate for the partner or keep perturbations manageable. In the antagonistic realm, opposite synergies are aimed at; accordingly, agents exploit glitches of the other person, and try to thwart or turn around their initiated dynamics. In the end, the synergy of the winning party succeeds, while that of the loser never materializes. Thirdly, some synergies are self-contained, as in many forms of dance co-improvisation where every moment stands for itself, while others cumulatively progress towards a goal, as in soft martial arts.[4] These typological considerations determine at what timescale and in what function to study synergy practices.

Overall, our qualitative micro-reconstruction captures the emergence of collective functions 'in the act', with attention to specific constituents and forms of interplay. It reveals the strategies whereby practitioners employ intra-body synergies to regulate inter-body synergies, what constraints and variations they exploit and what principles they bring to bear. To scholars, a micro-analytic perspective of this sort complements aggregate synergy measures by highlighting micro-aspects of synergy evolution and the underlying skills. This can benefit interaction science, biomechanics and sports psychology; among other things it can inform the 'frontloading' of phenomenological insights into quantitative designs and cross-validation through mixed methods approaches. Furthermore, since our reconstruction builds on the subjective meanings of practitioners, it lends itself to didactic applications, notably systematic formats for trainer–learner conversations, tools for practitioner self-reflection and several kinds of translational research.[5]

Notes

1 Mathematically, this implies that the behaviour of several agents can be expressed in a single equation of their coupling dynamics (*low-dimensionality*).
2 This requires tracking interpersonal 'variables that emerge from the interactions' (Passos et al. 2013, 1). There is a particular sensory range in a dimension such as distance, angle, alignment or speed difference that signals the affordance and frequently also informs about its optimality or not.
3 In collaborative settings contingencies are largely complementation-based, but may be of other kinds too, such as calculated perturbation or challenge.
4 A similar causal progression is typical of bodywork therapies, where clients receive a series of inputs for a combined synergistic health effect.
5 We wish to thank John Sutton, Kath Bicknell and our anonymous reviewers for their excellent suggestions for improving this contribution. We also want to express our gratitude to the dance, martial arts and acrobatics experts for their participation and the generous sharing of their insights. Finally, thanks goes to Hella Ebel and Loni Liebermann for letting us use their video material. This research was supported by research grants P-27870 and P-33289 from the Austria Science Fund (FWF).

References

Araújo, D. and K. Davids (2011), 'Embodied Cognition and Emergent Decision-Making in Dynamical Movement Systems', *Junctures: The Journal for Thematic Dialogue* 2: 45–56.

Araújo, D., and K. Davids (2016), 'Team Synergies in Sport: Theory and Measures', *Frontiers in Psychology* 7: 1449. https://doi.org/10.3389/fpsyg.2016.01449.

Bell, B., and S. Kozlowski (2002), 'A Typology of Virtual Teams: Implications for Effective Leadership', *Group & Organization Management* 27 (1): 14–49.

Bourbousson, J., C. Sève, and T. McGarry (2010), 'Space–Time Coordination Dynamics in Basketball: Part 2. The Interaction between the Two Teams', *Journal of Sports Sciences* 28 (3): 349–58.

Di Paolo, E., and H. De Jaegher (2007), 'Participatory Sense-Making: An Enactive Approach to Social Cognition', *Phenomenology and the Cognitive Sciences* 6 (4): 485–507.

Duarte, R., D. Araújo, V. Correia and K. Davids (2012), 'Sports Teams as Superorganisms: Implications of Sociobiological Models of Behaviour for Research and Practice in Team Sports Performance Analysis', *Sports Medicine* 42 (8): 633–42.

Froese, T., and T. Fuchs (2012), 'The Extended Body: A Case Study in the Neurophenomenology of Social Interaction', *Phenomenology and the Cognitive Sciences* 11 (2): 205–35.

Gibson, J. (1979), *The Ecological Approach to Visual Perception*. Boston: Houghton Mifflin.

Harrison, S., and M. Richardson (2009), 'Horsing around: Spontaneous Four-Legged Coordination', *Journal of Motor Behavior* 41 (6): 519–24.

Kimmel, M. (2021), 'The Micro-Genesis of Interpersonal Synergy – Insights from Improvised Dance Duets', *Ecological Psychology* 33 (2): 106–45.

Kimmel, M. (2017), 'The Complexity of Skillscapes: Skill Sets, Synergies, and Meta-Regulation in Joint Embodied Improvisation', In J. Gore and P. Ward (eds), *Proceedings of the 13th International Conference on Naturalistic Decision Making*, 20–23 June 2017, 102–9.

Kimmel, M. (2019), 'A Cognitive Theory of Joint Improvisation: The Case of Tango Argentino', In V. Midgelow (ed.), *The Oxford Handbook of Improvisation in Dance*, 562–92, Oxford: Oxford University Press.

Kimmel, M., and D. Hristova (2021), 'The Micro-genesis of Improvisational Co-creation', *Creativity Research Journal*. https://doi.org/10.1080/10400419.2021.1922197.

Kimmel, M., C. Irran and M. Luger (2015), 'Bodywork as Systemic and Inter-Enactive Competence: Participatory Process Management in Feldenkrais Method® & Zen Shiatsu', *Frontiers in Psychology* 5: 1424.

Kimmel, M., and C. Rogler (2018), 'Affordances in Interaction – the Case of Aikido', *Ecological Psychology* 30 (3): 195–223.

Kimmel, M., and C. Rogler (2019), 'The Anatomy of Antagonistic Coregulation: Emergent Coordination, Path Dependency, and the Interplay of Parameters in Aikido', *Human Movement Science* 63: 231–53.

Latash, M. (2008), *Synergy*. Oxford: Oxford University Press.

Passos, P., J. Milho and C. Button (2018), 'Quantifying Synergies in two-versus-one Situations in Team Sports: An Example from Rugby Union', *Behavior Research Methods* 50 (2): 620–9.

Passos, P., D. Araújo, and K. Davids (2013), 'Self-Organization Processes in Field-Invasion Team Sports: Implications for Leadership', *Sports Medicine* 43 (1): 1–7.
Passos, P., K. Davids and J. Y. Chow (Eds.) (2016). *Interpersonal Coordination and Performance in Social Systems*. Abingdon: Routledge.
Richardson, M., A. Washburn, S. Harrison and R. Kallen (2016), 'Symmetry and the Dynamics of Interpersonal Coordination', In Pedro Passos, K. Davids, and J. Y. Chow (eds), *Interpersonal Coordination and Performance in Social Systems*, 83–100, Abingdon: Routledge.
Riley, M., M. Richardson, K. Shockley and V. Ramenzoni (2011), 'Interpersonal Synergies', *Frontiers in Psychology* 2. https://doi.org/10.3389/fpsyg.2011.00038.
Schmidt, R., P. Fitzpatrick, R. Caron and J. Mergeche (2011), 'Understanding Social Motor Coordination', *Human Movement Science* 30 (5): 834–45.
Schmidt, R., C. Carello and M. Turvey (1990), 'Phase Transitions and Critical Fluctuations in the Visual Coordination of Rhythmic Movements between People', *Journal of Experimental Psychology: Human Perception and Performance* 16 (2): 227–47.
Torrents, C., R. Hristovski, J. Coterón and A. Ric (2016), 'Interpersonal Coordination in Contact Improvisation Dance', In P. Passos, K. Davids, and J. Chow (eds), *Interpersonal Coordination and Performance in Social Systems*, 94–108, London: Routledge.
Turvey, M (2007), 'Action and Perception at the Level of Synergies', *Human Movement Science* 26 (4): 657–97.
Turvey, M. T., R. Shaw and W. Mace (1978), 'Issues in the Theory of Action. Degrees of Freedom, Coordinative Structures and Coalitions', In J. Requin (ed.), *Attention and Performance VII*, 557–95, Hillsdale, NJ: Lawrence Erlbaum Associates.

Commentary: Mixing methods in the study of human action

Anthony Chemero

Back in 2013, Mike Richardson, Rachel Kallen (two of the authors of Chapter 8) and I used to talk a lot about how research on complex systems in the cognitive sciences focused too closely on artificial tasks. How often, we asked, do people wag their fingers along with a metronome? How often do they tap along with a metronome? How often do they say /ta/ once a second for twenty minutes? These phenomena are all naturally modelled using the methods of complex systems because they are all varieties of oscillation. Humans do many intelligent things on a day-to-day basis that can be understood as oscillating, but it is also very clear that oscillating constitutes a fairly small minority of our intelligent behaviour. Our goal, back then, was to use complex systems to model intelligent human phenomena that are not oscillations. To facilitate this, we, along with our then-PhD student Patrick Nalepka, developed a research project on virtual, social sheep herding. In the experiments, based loosely on earlier work by Dobri Dotov (Dotov et al. 2010), pairs of participants stood across from one another, on either side of a computer monitor projected onto to a glass table top. Their task was to 'herd' a collection of uncooperative balls ('sheep') to a circle ('pen') in the centre of the monitor. The participants were not allowed to talk to one another. Here, finally, was a task that had nothing to do with oscillating. Or so we thought. It turned out that the best strategy for containing the sheep was for the participants to repeatedly move the handheld controllers in a semi-circular pattern around borders of the pen in time with their partner. That is, they oscillated together (Nalepka et al. 2017, 2019). This is an interesting finding, of course, even though it was not what we expected.

There are two points to this story. The first is that oscillation is built deeply into human behaviour, but does not exhaust it. The second is that this set of experiments was an attempt to apply complex systems models to somewhat more naturalistic activities. We were not the only ones trying to do this. Alexandra Paxton, Drew Abney, Chris Kello and Rick Dale at University of California Merced (Abney et al. 2014) were starting to use complex systems models to study human dialogue. Ashley Walton, another PhD student that Richardson and I shared, was using complex systems models to study improvising jazz piano duos (Walton et al. 2015, 2018). Progress was being made. All of this is preface for examining a sort of tension that can be seen in Chapters 8, 9 and 10 of this volume. I will describe each chapter briefly before coming back to the tension.

The chapter by Kallen, Macpherson, Miles and Richardson is a plea to understand social interaction in terms of symmetry and symmetry breaking. These very general and very abstract concepts are typically spelled out in terms of Group Theory, but Kallen et al. spare us the technical details. They do provide a helpful example of how symmetry and symmetry breaking are relevant in social interactions. Imagine a round table set so that every seat is equally close to a glass to the left and a glass to the right. This is a perfectly symmetrical situation in which, from every seat, the glass to the left and the glass to the right seem equally for the person at that seat and no one knows which glass to use. The symmetry is broken as soon as someone takes a glass – once this happens, everyone else know which glass is theirs. The point here is that in perfectly symmetrical circumstances nothing happens; only when symmetries are broken do we get behaviour. Most symmetries are not perfect like this, and the degree of symmetry in a system can be quantified. Many useful tools for studying social interactions follow, allowing Kallen et al. to avoid the traps of 'either: (i) reducing phenomena to overly parsimonious mathematical formalisms (i.e. computational rules); or (ii) merely describing phenomena using the terminology of complex systems without any analytical rigor.' The tension I noted above stems from the fact that although the following two chapters certainly concern the detailed specifics of complex real-life human activities, they might at first glance seem to fall into Kallen et al.'s second trap of being 'merely descriptive' and 'lacking analytical rigor'.

Hjortborg's chapter is a detailed analysis of a single, expert elbow strike in a muay thai prize fight. The strike was executed by Singpayak ('Sing'), who is also a coach at the muay thai gym that Hjortborg joined as part of her research. (Hjortborg herself was already an experienced martial arts practitioner when she joined the gym.) The chapter includes a link to the fight, so readers can see this lightning fast, punishing elbow strike, which opens a gash on the forehead of Sing's opponent. In her interview with Sing, conducted while they watched the video of the fight together, Sing explains the strategy that created the opening for the elbow strike. Leading up to the strike Sing attacked repeatedly with 'push kicks', waist-high front thrust kicks, which disrupt his opponent's balance; to respond to these kicks, without giving ground, the opponent leans slightly forward, and, to block them, lower his hands. Just before the devastating elbow strike, Sing subtly fakes a push kick – so subtly that Hjortborg only noticed after Sing showed her while they watched the video – leading his opponent to lean forward and drop his hands, exposing his head. This was a strategy that Sing decided upon before going into the round; deciding upon this strategy constrained Sing's actions in the ring, enabling the cognitive activity that Hjortborg identifies in Sing's performance, which she refers to as 'low-commitment strategizing'. This low-commitment strategizing, Hjortborg argues, is *thinking*, and as such unaccounted for in Gibsonian (1979) and dynamical accounts (Silva et al. 2013) that focus on the tight connections between perception and action, and leave 'no room for thinking or cognition in action regulation'. Although Hjortborg's qualitative analysis occasionally traffics in complex systems terminology and employs no mathematics, it is hardly merely descriptive or lacking in analytical rigour.

Where Hjortborg dips her toes into complex systems terminology, Kimmel and Schneider dive in head first, casting their entire chapter in terms of the complex systems notion 'interpersonal synergy', which occurs when 'elements from different individuals enter into an interdependent organisation, establishing a macro-system with some collective functionality.' (This definition follows Araújo and Davids (2016).) What is novel, and especially valuable, about the chapter is that it develops a *qualitative* approach to interpersonal synergies, using all of the complex systems terminology, but none of the math. They go through a series of examples, including acroyoga, aikido and tango, exploring the ways participants in these activities participate in interpersonal synergies, or as they often put it, 'intercorporeal' synergies. In close interaction, components of the bodies of the interacting agents form synergies, forming a macro-system with a life of its own (i.e. with its own dynamics and functions that constrain the components of the individuals that make the system up). Aikido, tango and acroyoga all depend on willingness to participate, and even the following of explicit rules that are known to and followed by the participants. An elbow strike to the forehead, allowed in professional muay thai bouts, would certainly disrupt the interpersonal synergies that form in tango or aikido. Even in less constrained activities, like muay thai or mixed martial arts, there are rules that participants follow, such as not striking one another in the groin. And crucially, the participants need not be willing participants in the intercorporeal synergies they become part of. In aikido, Kimmel and Schneider show, combatants each attempt to direct the interpersonal synergy that forms to their opposing advantages. Here, still, participants have agreed to the rules of the situation, but that is not always the case: sometimes you simply cannot avoid the aggressive or extremely intoxicated person that confronts you, and you work to alter the interpersonal synergy to your favour (i.e. to end the interaction).

Here, finally, we can see the tension: Kallen et al. urge against research without analytical rigour. Yet I don't doubt for a second that they would be impressed by the purely qualitative work and extraordinarily insightful work described in the chapters by Hjortborg and Kimmel and Schneider. The conclusions they draw from their qualitative studies of complex human actions are unavailable from a quantitative perspective, which simply has nothing to say about strategies (at whatever level of commitment) or rule following. The differences in the insights available to qualitative versus quantitative analyses align closely with the old philosophical distinction between actions and happenings. To see the distinction, consider the difference between winking to send a friend an unspoken message and blinking because of some dust in the eye. Even though the fluttering of the eyelid might be identical in these two cases, they are not the same. The winking is suffused with purpose and intention; it is subject to norms. None of this applies to the blink (Ryle 1979). Qualitative analyses give us access to events as actions, while quantitative analyses give us access to events as happenings. None of this implies that there are two things going on when I wink. As Ryle (1979) put it, a penny is not nothing but a metal disc, but it is also not something else in addition to a metal disc. In certain contexts, we can consider the penny as a metal disc, as when we weigh it or calculate the force of the impact on the floor when it falls off the table. In other contexts, we can consider the penny as currency, as something that, in certain contexts, can be exchanged for goods or services. Here, I am disagreeing with Hjortborg's claim that Gibsonian and perception–action views leave no room for thinking. She argues that her analysis of Sing as employing low-commitment strategies implies that there is thinking and cognition, in addition to the control of action. I would suggest instead that understanding Sing as engaging in low-commitment strategizing is to consider his movements around the ring as actions, rather than just happenings. As such, those actions are normatively evaluable and imbued with intention and purpose. But they are not another thing, in addition to his movements.

The conclusion I wish to draw from this is that qualitative and quantitative approaches need one another, especially in the cognitive sciences. It is crucial to know, in detail and with precision, what happens (e.g. how the eyelid moved); it is equally crucial to know, in context, what the people are doing (e.g. whether the blink's message was received). This echoes a point I have made in other contexts (e.g. Chemero 2009; Baggs and Chemero 2021): the dynamical approach is extraordinarily valuable for making sense of data, but, to be complete as a psychology, it needs to be supplemented with a story about agency. I realize that dealing with quantitative and qualitative data are advanced arts, and mostly operate separately from one another. This is why mixed methods research is so deucedly difficult. And to be clear, I am not suggesting that the authors of Chapters 9 and 10 learn a bunch of calculus, or that the authors of Chapter 8 learn how to run focus groups or do interviews. I am calling, instead, for collaboration: for the sequel to this book, I hope to see a chapter co-authored by some of the authors of these chapters in which they apply both qualitative and quantitative methods to the same set of participants and activities.

References

Abney, D. H., A. Paxton, R. Dale and C. T. Kello (2014), 'Complexity Matching in Dyadic Conversation', *Journal of Experimental Psychology: General 143* (6): 2304.

Araújo, D. and K. Davids (2016), 'Team Synergies in Sport: Theory and Measures', *Frontiers in Psychology* 7: 1449.

Baggs, E., and A. Chemero (2021), 'Radical Embodiment in Two Directions, *Synthese* 198 (9), 2175-90.

Chemero, A. (2009), *Radical Embodied Cognitive Science*. Cambridge, MA: MIT Press.

Dotov, D. G., L. Nie and A. Chemero (2010), 'A Demonstration of the Transition from Ready-to-hand to Unready-to-hand', *PLoS One* 5 (3): e9433.

Gibson, J. J. (1979), *The Ecological Approach to Visual Perception*. Boston: Houghton Mifflin.

Nalepka, P., R. W. Kallen, A. Chemero, E. Saltzman and M. J. Richardson (2017), 'Herd Those Sheep: Emergent Multiagent Coordination and Behavioral-Mode Switching', *Psychological Science* 28 (5): 630–50.

Nalepka, P., M. Lamb, R. W. Kallen, K. Shockley, A. Chemero, E. Saltzman and M. J. Richardson (2019), 'Human Social Motor Solutions for Human–Machine Interaction in Dynamical Task Contexts', *Proceedings of the National Academy of Sciences*, 116 (4): 1437–46.

Ryle, G. (1979), *On Thinking*. K. Kolenda (ed.), Oxford: Blackwell.

Silva, P., J. Garganta, D. Araújo, K.Davids and P. Aguiar (2013), 'Shared Knowledge or Shared Affordances? Insights from an Ecological Dynamics Approach to Team Coordination in Sports', *Sports Medicine* 43 (9): 765–72.

Walton, A. E., M. J. Richardson, P. Langland-Hassan and A. Chemero (2015), 'Improvisation and the Self-Organization of Multiple Musical Bodies', *Frontiers in Psychology* 6: 313.

Walton, A. E., A. Washburn, P. Langland-Hassan, A. Chemero, H. Kloos, and M. J. Richardson (2018), 'Creating Time: Social Collaboration in Music Improvisation', *Topics in Cognitive Science* 10 (1): 95–119.

Afterwords

Commentary: Ecologies of acting and enacting

Catherine J. Stevens

This volume is rich with accounts of human performance, expertise, techniques to support learning, methods of practice and rehearsal, and the associated agonies and ecstasies of adults striving to break habits, acquire and refine new habits, and excel in high stakes physical environments. Each account demonstrates the interconnection of action and thought.

In scientific theories of human cognition, action and thought have been termed, respectively, procedural knowledge (knowing how and often seen as unavailable to conscious awareness) and declarative knowledge (knowing what, able to be declared, thus available to conscious awareness).

In this commentary, I discuss the interplay of procedural and declarative knowledge in skill acquisition and deployment – knowledge in doing. I also touch on the dynamics of control in skill ecosystems and egalitarianism associated with a blurring of performer and audience. Finally, I consider the role of multiple modalities as cues, and their benefits, for meaning making, memory and retrieval.

Knowledge in doing

Skill acquisition, whether in the context of elite performance such as dance or music or defying the limits of physicality in acrobatics and gymnastics, freediving or martial arts, involves 4E cognition. That is, cognition is *enactive*, *embedded* in an environment and physical, social and political context, *embodied* in one's physical and personal form, and *extended* beyond the brain and skull to tools, artefacts, peripersonal space, the environment and other people. We see 4E cognition in Tribble's analysis of Shakespeare's Globe theatre, in Bicknell and Brümmer's handstands, in Downey's description of freediving, in Pini's account of Body Weather, in the synergy logic of martial and bodily arts described by Kimmel and Schneider, and in the use of an artificial neural network by musician Holly Herndon. Mastering freediving or composing music or dance, for example, engages knowing how and knowing what. The two systems interact as skills are acquired and consolidation in memory takes place. In these systems, action is entwined with thought.

A compelling example of action entwined with thought is that of Scrabble players and their use of letter tiles (Maglio et al. 1999). Physical manipulation of task-relevant objects, rather than simply observing them, can aid problem solving. In other words, physical activity complements cognitive activity. Physical interaction can also broaden the basis of creativity (Smithwick and Kirsh 2015). In another study, architects were asked to use blocks to design their dream house. Architects compared with novice participants used a generative rather than representational approach. For example, novice participants used blocks to represent structures such as an entrance. Architects moved shapes around to explore relationships between blocks and the site. Architects imagined spaces being occupied in various ways. The experts' physical manipulation technique is visual-kinaesthetic experimentation (Smithwick and Kirsh 2015, 2239). Smithwick and Kirsh's interpretation is that the expert's physical interaction with the blocks increases the dimensionality of the design space – physical exploration facilitates 'perceptual plurality' with shapes being 'relentlessly ambiguous' (p. 2239). The diversity of appearance of the blocks supports 'indefinitely many semantics'. For example, in one semantics, the focus may be 2D shapes visible from volumes, in another, the vertices carry meaning. In the architects' use of the blocks there can be a massing semantics map, a negative space semantics map and so on. The blocks give rise to a multiplicity of meanings and interpretations. The design space with blocks is expanded with perceptual richness and unlimited ambiguity (p. 2240). Here, as in Martens and Rietveld's chapter on architecture, the physical supports the conceptual countering the assumption that design is higher creative cognition that first happens in the individual brain. The physically demanding handstands described by Bicknell and Brümmer flip the situation; their play with and use of concepts and language helps to constrain, and decrease the dimensionality of the action space or task. The phrase 'we must be elephants' chunks seconds of physics, visual flow, proprioception and kinaesthesia into a phrase. This additional mode labels a new motor habit and skill. Such vocabulary is also conveniently asynchronous with action.

A dynamical system of perception and action, Bicknell and Brümmer's physicality and actions are associated with semantics that disambiguate and help to simplify multiple physical, spatial and temporal relations. Spoken language chunks and codes; it is shareable, communicable and social. The two learners – learning companions – bond through humour, language and metaphor. They lighten the burden of trial and error with laughter and imagery. Strategically, their second reference to elephants captures a shared goal in positive, hopeful language. The interconnectedness of action and thought is palpable – there is knowledge in doing.

The dynamics of control and inclusion

In her analysis of Shakespeare's Globe theatre, Tribble describes actors relating to one another and anticipating and responding to cues, props and narrative. Actors also interact with audience members, one-off and sometimes emergency situations, in the moment. The notion of a set work begins to dissolve and, so too, the distinction between performer and audience, real life and play. The ecosystem is again enactive, embedded, embodied and extended.

Shakespeare's Globe theatre is a dynamical system with elements in the system shaped by, and shaping, each other (Mitchell 1998). Control is at times centralized and at times distributed (Heylighen 2001). Tribble's analysis including Front-of-House makes clear how much control is at the heart of live performance. For example, managing and admitting the audience, permitting intermission, managing problems such as audience illness or disruption from cell phone use, talking and so on. Open fire, artificial oil lighting and later introduction of gas and then brighter incandescent lights changed the context for live performance (e.g. Essig 2007; Goron 2016) with the audience, for example, controlled through being in the dark relative to the performers. At the Globe, things are more egalitarian with actors and audience in shared light, both being able to influence the performance, and with audience disruptions at times being integrated into the show.

Roberts and Krueger discuss agency in art, composition and performance. Why is *Spawn*, an artificial neural network for creative purposes, unsettling? Is it a loss of control and/or the absence of a human relationship in the creative process? In experiments with contemporary dance artists, we have demonstrated the positive association with dancers, for example, improvising together compared with alone. Interacting with others increases the possibilities, risks and play. While the quantity of ideas may not differ in pair and trio improvisation compared with solo improvisation, the quality of the experience changes (Leach and Stevens 2020; Stevens and Leach 2015). Recall from long-term memory for set contemporary dance phrases is also greater when the phrase is performed in a duo compared with a solo (Stevens et al. 2019).

Roberts and Krueger resolve the question of AI as agent by considering Holly Herndon's treatment of the artificial neural network *Spawn* as fictional – treating *Spawn* 'as if' it is creative. Thus, *Spawn* is more than a sounding board from which to riff. Stimuli, boundary spanners, in the form of other people, events, locations, objects, works of art, textures, smells and so on, can inspire/drive creative processes (Sawyer 2012; Stevens et al. 2003). The further step is that creativity emanates through engaging human imagination and 'as if' fiction. Problem finding as well as solving, generating and exploring (Finke et al. 1992), play and discovery, are elicited in treating *Spawn* as if it is a creative partner.

Empirical analysis of the 'as if' hypothesis could adapt the Turing Test for an observer's belief in 'as ifness'. In the Turing Test, the maxim that computers can think is tested with an interlocuter probing a computer and a human with identical questions to see if the computer can be distinguished from the human. One could develop a similar test to compare Herndon's treatment of an agent as if it is creative with treatment of an agent that is creative (e.g. another composer). The test could take the form of an observer challenged to distinguish the composer's interaction with a creative, compared with a non-creative, agent.

Alternatively, we could scrutinize the creative process itself. Design theory, for example, depicts idea generation as following either an iterative process or a 'fail fast, fail often' process. We analysed the time course of creativity in the contemporary dance company, Australian Dance Theatre (Kirsh et al. 2020). Seven professional dance artists were asked to improvise movement material to make a new dance phrase, in response to choreographic tasks (Forsythe 1999) in fifteen-, thirty- and forty-five-minute time periods. We observed a pattern where ideas are created and aggressively pruned early, and with many early ideas making it into the final product or phrase. The results reflected a blend of the iterative and the fail fast fail often approaches. To what extent does the creative process with *Spawn* align with documented accounts of human creativity?

Multimodal cognition in action and benefits of redundancies

What are the long-term benefits of many modalities, such as language, motor control and stimulation of six sensory systems, for skill acquisition, performance and the interconnectedness of action and thought? From a practical point of view there are likely benefits for learning, retention and retrieval when there are redundancies in a system. Multiple modalities – spatial, temporal, verbal, kinaesthetic, proprioceptive, auditory, olfactory, visual – give rise to multiple layers of meaning and perceptual plurality. Layers of meaning give depth and many facets to ideas and expression. The one-to-many mappings, for example, of language in poetry, or the non-referential nature of instrumental or laptop music, or movement for its own sake in contemporary dance, are quintessential in Western art. Across repeated viewings or listening, different layers, experiences, ideas, feelings may arise. There can be deliciously playful or political ambiguities. There may be fleeting synaesthesia (connections across senses) when ideas and thoughts are generated or declared through sound, action, movement and stillness and shades of light and dark.

Redundancy in skill ecosystems might be an insurance policy from the perspective of human learning and memory. When so many modalities have been engaged in learning, rehearsal and reproduction, stimulation of one modality or a particular cue, by spread of activation (e.g. Collins and Quillian 1970), can stimulate other modalities. Stevens, Ginsborg and Lester (2011) reported the multiple modes recalled from long-term memory for dance ranging from the politics of the day when the dance exercise and movements were first learned through to the accompanying music and to the smell of the hall. 'No elephants today' (Bicknell and Brümmer) is a multifaceted, architectonic cue for preparation, execution and completion of a handstand. The synergy logic (Kimmel and Schneider) captures the inter-dependent unfolding physics of two or more martial artists. Body Weather (Pini) encompasses all perceptual and cognitive systems.

Multiple and overlapping modalities provide a range of cues for retaining and reproducing complex actions. Layers of meaning and perceptual plurality enable varied entry points that can optimize the diversity of audience inclusion and engagement. There are profound implications for learning, education and creativity as we recognize cognition in and through action experiences.

References

Collins, A. M. and M. R. Quillian (1970), 'Facilitating Retrieval from Semantic Memory: The Effect of Repeating Part of an Inference', *Acta Psychologica 33*: 304–14.
Essig, L. (2007), 'A Primer for the History of Stage Lighting', *Performing Arts Resources*, 25: 1–17.
Finke, R. A., T. M. Ward and S. M. Smith (1992), *Creative Cognition: Theory, Research, and Applications*. Cambridge, MA: MIT Press.
Forsythe, W. (1999), *Improvisation Technologies: A Tool for the Analytical Dance Eye*. Karlsruhe: Zentrum für Kunst und Medientechnologie.
Goron, M. (2016), '*Patience* at the Savoy', In M. Goron, *Gilbert and Sullivan's 'Respectable Capers'*, 73–106. London: Palgrave Macmillan.
Heylighen, F. (2001), 'The Science of Self-Organization and Adaptivity', In *The Encyclopedia of Life Support Systems*, 253–80, Oxford: EOLSS Publishers Co. Ltd.
Kirsh D., C. J. Stevens and D. W. Piepers (2020), 'Time Course of Creativity in Dance', *Frontiers in Psychology 11*: 518248.

Leach, J. and C. J. Stevens (2020), 'Relational Creativity and Improvisation in Contemporary Dance', *Interdisciplinary Science Reviews* 45 (1): 95–116.

Maglio, P., T. Matlock, D. Raphaely, B. Chernicky and D. Kirsh (1999), 'Interactive Skill in Scrabble', In *Proceedings of the 21st Annual Conference of the Cognitive Science Society*, Mahwah, NJ: Lawrence Erlbaum.

Mitchell, M. (1998), 'A Complex-Systems Perspective on the 'Computation vs. Dynamics' Debate in Cognitive Science', In M. A. Gernsbacher and S. J. Derry (eds), *Proceedings of the 20th Annual Conference of the Cognitive Science Society*. Hillsdale, NJ: Erlbaum.

Sawyer, R. K., (Ed.) (2012), *Explaining Creativity: The Science of Human Innovation*. 2nd ed. Oxford: Oxford University Press.

Smithwick, D. and D. Kirsh (2015), 'Let's Get Physical: Thinking with Things in Architectural Design', In Noelle, D. C., Dale, R., Warlaumont, A. S., Yoshimi, J., Matlock, T., Jennings, C. D., and Maglio, P. P. (eds), *Proceedings of the 37th Annual Meeting of the Cognitive Science Society*. Austin, TX: Cognitive Science Society.

Stevens, C., J. Ginsborg and G. Lester (2011), 'Backwards and Forwards in Space and Time: Recalling Dance Movement from Long-term Memory', *Memory Studies* 4: 234–50.

Stevens, C. J. and J. Leach (2015), 'Bodystorming: Effects of Collaboration and Familiarity on Improvising Contemporary Dance', *Cognitive Processing – International Quarterly of Cognitive Science* 16 (Suppl 1): S403–S407.

Stevens, C., S. Malloch, S. McKechnie and N. Steven (2003), 'Choreographic Cognition: The Time-course and Phenomenology of Creating a Dance', *Pragmatics & Cognition* 11 (2): 297–326.

Stevens, C. J., K. Vincs, S. deLahunta and E. Old (2019), 'Long-Term Memory for Contemporary Dance is Distributed and Collaborative', *Acta Psychologica* 194: 17–27.

Commentary: Betwixt and between

Ian Maxwell

Perhaps the most exciting aspect of being involved in the work that has led to this collection of essays is that it has been an iterative, collaborative process. I will not use the constrained but generous space I have been offered here to attempt to synthesize, extract key outcomes from or draw some kind of conclusion from the dense, deeply engaging polyvocality of the essays in this volume. Rather, I want to take the opportunity to reflect on that process and to think about what the process allows us to understand about interdisciplinarity.

To briefly recall the terms of that process: I was invited by John and Kath, early in 2020, to join a series of workshops. I believe these were originally conceived as taking place by some kind of video conferencing, so as to afford the opportunity for trans-hemispheric engagement. By the time we reached August, the generic notion of video conference had been displaced by the ubiquity of Zoom. We met during the unsettling onset of COVID-19, constellating a community of co-investigators all-too-pleased to be exploring the affordances of what was becoming the new familiar technology and its potential to overcome the logics and constraints of 'social distancing'. In clusters of three, colleagues presented drafts of the papers which find their final form in the pages of this volume. There was no compulsion to present finished work. Each contributor was assigned two respondents, who, after the presentation, were invited 'to "riff" to and with the work presented': to prod, provoke, take up, extend, think with and beyond that work. Later, those of us framed as respondents – 'commentators' – were given the opportunity to read more finished papers, and to provide written responses on some of them as part of a review process, before, ourselves, pulling together these short written pieces, pitched at to allowing us 'to consider and develop themes or topics from as many or as few of the individual chapters as you like'.

One of my thoughts is that what is of value in such a project is, to no small degree, the manner in which it *resists* synthesis, outcomes and conclusion. That is, rather than settling anything, the main legacy of this work – of which this volume is only a trace – is an *un*settling: a disturbance of the confidences, assurances and certainties which are part of the powerful and, most frequently, the desirable outcomes of disciplinarity. This is not intended as a repudiation of disciplinarity, nor yet as an endorsement of a kind of laissez-faire embrace of *inter*disciplinarity. Rather, it is to take a brief moment to reflect on the work *behind* this volume, and to wonder about what it shows us.

Disciplines discipline. That is, they create (relative) stabilities – perhaps closures, or at the very least, contingent constraints upon what Derrida characterized as the 'joyous affirmation' of interpretation as playful creativity (1967, 289) – amidst the openness of our attempts to understanding aspects of our being in, and our experiences of, the world. They shape our thinking, our practices, framing the questions we ask, the objects we constitute to ask questions about. A corollary of this, of course, is that disciplines *exclude* and *obscure*: in foregrounding some phenomena, disciplines push other phenomena to the side. They afford and constrain how we constitute knowledge and what knowledge is. That is, they are *technés*, in the Foucauldian sense, functioning both – often simultaneously, and always to various degrees – to exert top-down governmentality, and to enable possibilities for creative empowerment.

Disciplines do not, then, simply define, divide or distinguish domains of knowledge – although that is part of what they do. They establish and maintain ways of doing things. Disciplines are social practices as much as they are constituted by abstracted principles: as epistemologies, methodologies and so on. Disciplines are constituted as logics of practices, shaped over time, institutionalized, challenged, disrupted, overturned as part of continuous contestation and struggle. Disciplines, as established, defended, elaborated, contested, institutionalized may usefully be thought of as genres of social practices. This is not to reduce disciplines to 'mere' sociality, but to note they are marked, special kinds of practices – *ecologies of skill*, to pick up one of the key themes with which this volume is concerned – not just 'ways of doing things'.

How, then, might we think of what *inter*disciplinarity is or might be? Does the movement implied in the prefix involve forging a kind of contract between sovereign domains? Or might it constitute a challenging of such sovereign notions? More: does it not, in some circumstances, enjoin an agonistics: a disputing of territory? A creation of hierarchies? A play of deterritorializing and reterritorializing? By the same token, the negotiation of difference can take the form of a collaboration: a playing *with*, rather than a playing *against*, even when that play is oriented not towards resolution, but towards seeing what the game is capable of.

Alternatively, might interdisciplinary practice entail a resolute commitment to taking up the space between disciplines, as a kind of interstitiality, involving, perhaps, a refusal to take up a stable disciplinary set of practices? Or confecting an assemblage of disciplines, a kind of Swiss Army Knife of practices, to be deployed in response to particular questions?

Writing of genres, Briggs and Bauman (1992) set off from a critique of what they characterize as 'top-down' models which constitute genres as (quasi-)scientific taxonomies, understanding generic categories as mutually exclusive. Against this, they set bottom-up models – ethnographic investigations of local genre systems – which reveal 'generic categories that overlap and interpenetrate in a range of complex ways, or aspects of verbal production that are resistant to orderly categorisation' (1992, 144). Their key interest is to interrogate

> the distinction between approaches to genre that on the one hand constitute genre as an orderly and ordering principle in the organization of language, society and culture, from those that contend with the elements of disjunction, ambiguity and general lack of fit that lurk around the margins of generic categories, systems and texts. (1992, 144–5)

Briggs and Bauman extend Bakhtin's reframing of literary texts as intertexts – the insight that literary structure does not simply *exist* but is generated in relation to *another* structure – to thinking about genre. For them, genre is quintessentially intertextual: genres 'exist' in relationship to other genres, distinguished and distinguishing, excluding and including, defining against and defining alongside. Boundaries may be tightly policed, or leaky, but are always porous.

More, Briggs and Bauman associate the establishing and maintenance of genres with the creation of 'imagined communities': the establishment and maintenance of orderly social systems. They identify two strategies by which these processes may unfold: one involves minimizing the gap between one's practice (they refer, here, to the performance of discourse) and the genre, 'rendering the discourse maximally interpretable through the use of generic precedents' (1992, 149). Such an approach 'sustains highly conservative, traditionalizing modes of creating textual authority' (1992, 149). Alternatively, emphasizing the gap between a given (discursive) performance and a genre 'underlies strategies for building authority through claims of individual creativity and innovation [...] resistance to the hegemonic structures associated with established genres, and other motives for distancing oneself from textual precedents' (1992, 149).

Disciplines qua genres, then, might be thought about in such terms. The various essays in this volume engage a range of strategies: some take up strong disciplinary voices, contributing to the interdisciplinary milieu – the ecology of the volume – from positions of relative stability; others enjoin an explicitly interdisciplinary collaboration, some in the form of co-authorship, others in terms of what Bakhtin would refer to as a heteroglossic intertextuality, in which a singular authorial voice itself negotiates the tensions and alignments between disciplinary practices. Many advance what amount to 'claims of [...] creativity and innovation', not necessarily in terms of 'individuality', as Briggs and Bauman have it, but as a group, movement or small team exploring novel possibilities.

Perhaps the most signal quality among the contributions is a willingness to complicate the question of disciplinarity in their own work, or, perhaps, to bring to the foreground the kinds of complications inherent in our relationships to disciplinarity: the kinds of complications to which Briggs and Bauman draw our attention. For some contributors, this is developed by means of an explicit collaboration: Bicknell and Brümmer, for example, perform, in their research and writing, a careful, sustained dialogue, while Martens, Rietveld and Rietveld model an expansive, multilateral (as it were) conversation between and across practices. For others – Downey, Tribble – the interdisciplinary problematics unfold through their reaching beyond their 'own' fields, drawing upon and speaking to fields other than their own, with all the risks attendant to such a project (how might, for example, the use of a term such as 'structure' resonate across disciplines?). At the other end of the continuum, there is work that might be characterized as developing multimodal approaches within relatively stable disciplinary sites (Kallen et al.), and work which enacts a disciplinary expansiveness by turning a strong disciplinary focus towards novel (from that disciplinary perspective) phenomena (Roberts and Krueger).

Going a bit further, Briggs and Bauman direct our attention not only to the manner in which we articulate our practice to disciplines, and not only to the relative porosity of disciplinary boundaries, but towards the very betweenness from and against which those boundaries take their form. Above, I raised, as a kind of mind experiment, the possibility of inhabiting that in-between space as, perhaps, a radical conception of interdisciplinarity as sustained interstitiality: a dwelling among, between, betwixt, in the midst of, disciplines. What might such a dwelling entail? Not just the nominal designation of a shelter or home *in* the in-between, perhaps but a sustained project of making (and unmaking/remaking) ways of being – of practices in and across those gaps. Indeed, a number of essays in the volume take up such an in-betweenness, perhaps precisely because their work grapples with the question of bodily in-betweenness: Pini's work on the radical encounters between bodies and environment in Body Weather practice; Ravn, Hjortborg, and Kimmel and Schneider on the intensity of intercorporeal encounters in martial arts and other embodied dyadic practices.

And here I play my own disciplinary hand, I perform the gesture that most readily comes to me, and which I found myself performing regularly throughout the process that shaped this collection. That is, the genre of the ethnographer, the figure who moves across and between cultural worlds or, in this instance, across and between the genres of practice we call disciplines. The conception of ethnographer I have in mind here is that described by Clifford Geertz: one who moves between cultural worlds but does so without making a claim towards totalizing knowledge; one who, instead, takes up the task of mediating between those worlds, a mediation conceived in terms of movement and translation (see Geertz 1974). The movement, however, is between, not over (or under): the Geertzian ethnographer at no point extracts themselves from the problematic of in-betweenness, is never able to take up either a panoptic overview with a view to forging a synthesis, nor attempting to look through (and under) difference in order to discern first principles.

The Geertzian ethnographer traverses in-betweenness, moving across the gaps between cultures, approaching difference without ever becoming the other; nor yet able to return entirely to from whence they came. The experience is transformative, disruptive and fundamentally mediative. To move in this place is, in a sense, to let go of an aspiration to solid foundations or to the possibility of epistemological closure: to forge a level of accommodation with dis-ease and contingency. It maintains, when taken up as a sustained practice (beyond the *strictu sensu* scene of anthropological methodology per se), something of an analogy to Brechtian notions of *Verfremdungseffekt* – the defamiliarizing effort to render that which is nearest to hand as *strange* or remarkable – or, perhaps, to the Husserlian effort to suspend the natural attitude in order to (re-)encounter the essence of our consciousness of the world, freed from the contingencies of 'common sense'.

As translator, the ethnographer constantly transposes questions of cultural practice from one context – one social genre – to another. This is challenge taken on – indeed, embraced with care and attention – by many of the authors in this volume. Predominantly, this takes the form of finding linguistic analogies or differences, in order to reframe 'insiderness' – perspectives on the world, styled, by Geertz, as *sensibilities* – in such a manner as to render that sensibility to a greater or lesser extent accessible to outsiders (recalling, too, that 'insiderness' and 'outsiderness' are not themselves absolute qualities, but are themselves governed by the principle of 'to a greater or lesser extent'). More, given that the 'object' of the ethnographer's inquiry is a *sensibility* – a feeling for life, as much as a set of propositions *about* life – the knowledge rendered by the ethnographer as language is always incomplete, partial and, again, to a greater or lesser extent, unreliable.

This is not, however, to give over to epistemological arbitrariness, nor to a kind of relativistic free-for-all, in which all that is stake is the assembling of examples, placed in colourful juxtaposition. For Geertz, what is at stake is the effort to establish and to sustain a conversation, not on the grounds of a shared set of principles or understandings, but on the far less stable, but nonetheless pragmatic, grounds of transaction and contingency. That is to say: research construed as an enduring struggle not only to understand aspects of the world, but to make ourselves understood to each other. In a real sense, then, the volume presents as an extended, collaborative project about the skills required for, and the performance of, the domain of interdisciplinary research – in all its complexity, and its capacity to uncover the world in novel, challenging and potentially unsettling ways.

References

Briggs, C. L. and R. Bauman (1992), 'Genre, Intertextuality and Social Power,' *Journal of Linguistic Anthropology 2* (2): 131–72.

Derrida, J. (1967), 'Structure, Sign, and Play in the Discourse of the Human Sciences', trans. A. Bass, and A. Bass (eds), (1978), *Writing And Difference*, 278–93, Chicago: University of Chicago Press.

Geertz, C. (1974), '"From the Native's Point of View": On the Nature of Anthropological Understanding', *The Bulletin of the American Academy of Arts and Sciences 28* (1): 26–45.

INDEX

acroyoga 191–2, 210
actors 23–32, 69–71, 217–18
  See also theatre
affordances 53–7, 61–5, 161, 195–7
agency, distributed or ecological 25, 35–47, 72, 133–5, 145, 218
  See also creativity, self-transformation, skill
agency, musical 125–8, 132–4, 145–6
aikido 12, 109–22, 144–5, 147, 193–4, 200, 210
air hunger 95–6, 100, 102–3
algorithms 126, 145–6
antagonistic interaction 109–10, 120–1, 144, 172–3, 179–83, 189, 195–8, 202
apnoea *See* breath, breathing
apprenticeship learning 11, 37, 110–15
Araújo, Duarte 162, 188, 210
architecture 12, 53–66, 70–1, 159, 216–17
art worlds 22–3, 28
artefacts *See* equipment
artificial intelligence (AI) 3, 6, 125–36, 145–7, 216, 218
assemblages, soft assembly 22, 95–6, 102–5, 187–90, 195, 198–200
attention, education of 37, 43–4, 84–5, 87, 97–8, 113, 117–18, 142–4, 178

attention, expert 4–5, 44, 116–18, 174, 177–9, 196–8
attention, management of 22–5, 30, 64, 70, 190
audiences 8, 13, 22–33, 36, 41–4, 71, 127, 132, 217–20
  See also listening
authorship, authority 27, 30, 70, 96, 133, 226
  See also status
AURA NOX ANIMA 37, 40–3, 46–8
automatic actions 3, 9, 11, 94–5, 101–2, 172–3, 181
awareness, bodily 37, 42–5, 70, 77–80, 84–90, 112–16, 142–4, 174–5, 183
  See also kinaesthesia, mindful performance
awareness, ecological 11, 37, 43–7, 72
awareness, free-floating 9, 11, 44, 81, 115, 183, 201

baby freeze 147–8
Bakhtin, Mikhail 226
beauty 71, 159–60
Becker, Howard S. 22–3, 28
Bicknell, Kath 89, 142–3, 146, 216–17, 220, 227
bisoku See stillness and slow movement
Body Weather 13, 36–47, 72, 216, 220, 227
bounds of cognition, the 7–8
Bourdieu, Pierre 80, 82, 87

INDEX

brains 6–7, 94, 100, 216–17
breakdance 147–8
breakdowns *See* glitches
breath, breathing 93–105, 118, 127, 143–4
Bresnahan, Ali 36
Brümmer, Kristina 142–3, 146, 216–17, 220, 227
Bunker 599 58, 61, 65, 71
Butoh 38–9, 48

Casey, Edward 45
choosing, choice 70, 157, 180–3
Christensen, Wayne 81, 87, 89, 172
Clark, Andy 36
coaching *see* instruction
cognition 1–4, 9, 54–6, 71, 81, 103–4, 127, 172–4, 181–3, 209–11, 219–20
 See also distributed cognition, 4E cognition, multimodal cognition
cognitive ecologies 4–8, 10–12, 21–3, 25–7, 32, 38, 45, 60–1, 70, 101–5, 178–9
cognitive humanities 5–8, 215–20
collaboration 1–5, 8–15, 21–8, 32, 55–66, 70–2, 82–90, 94–105, 125–37, 141–9, 187–202, 223–9
 See also interdependence, joint action
composition 126–34, 145, 160, 216, 218
concrete 58, 66, 71
Contact Improvisation 199–201
coordination 22–32, 54–7, 119–21, 153–66, 180, 187–9, 193–4, 198–9
cos-players, Tudor 29
COVID–19 141–2, 147–8, 223
creativity 10–12, 21–2, 32–3, 38, 41–3, 125–38, 145–6, 159–60, 188, 199–200, 216–20, 224–6

crocodiles (experienced aikido practitioners) 115–21
cues 4, 10, 81–7, 97, 217, 220
 See also multimodal cognition, nudges
cultural history 58–66

dance, dancers 35–48, 72, 120, 147–8, 158, 161, 193–4, 198–200, 202, 218–20
Davids, Keith 162, 188, 210
de Jaegher, Hanne 119–20
de Quincey, Tess 39, 42–4
Deltawerk // 58–9
DeNora, Tia 134
Derrida, Jacques 224
design 12, 24, 28, 135, 216–19
DeZutter, Stacey 23, 32
diagnosing 83, 86, 172, 177–83
disciplines, disciplinarity 224–9
 See also interdisciplinarity
distributed cognition 6–7, 10–12, 21–2, 25, 45–6, 90, 102, 132, 178, 217
 See also cognition, 4E cognition
dive reflex 100–2
diving, divers *See* freediving
division of labour 6, 9–10, 192–3
Downey, Greg 85, 115, 141, 143, 146, 149
Dreyfus, Hubert 3, 94
dunes 13, 37, 40–5, 72

Eccles, David W. 173
ecological dynamics 21, 54, 62, 153–5, 162, 182, 209–11
ecologies of skill 1–5, 8–15, 21–2, 25–8, 54, 78–9, 103–5, 110–15, 121, 143–5, 173–5, 187–202, 215–20, 225
embodied practices 2–6, 8–12, 37–8, 45–7, 57–8, 64–5, 79–82, 88–9, 95–105, 111–21, 126–7, 141–9, 171–83, 187–202
 See also skill

INDEX

emergence 9–10, 21–3, 32, 154–66, 199–202
emotion 2–6, 86–8, 97, 126–8, 159–60, 164, 174
Eno, Brian 130, 136
equipment 4, 10, 22, 63, 69, 83–4, 93–4, 100, 102, 104, 126–9, 137, 148, 160, 216–17
See also sociomaterial systems
Eterna, Lux 40–7
ethnography 2, 7, 11, 15, 56, 58, 78, 82–3, 85, 89–90, 110–11, 141, 173–5, 178, 183, 225–9
experience-near case studies 2, 7–8, 11–12, 103, 183
See also ethnography
experimental psychology 7, 82, 144, 164, 189, 207–12
expertise *See* skill

failure 77–90, 142
fakes, feints, lures, deception 175–7, 195–8, 209
See also antagonistic interaction
falls, falling 109–10, 143
fiction, fictionalism 128–37, 145–6, 218
See also make-believe
fights, fighting 109, 144, 171–83, 209
flying fish 29
4E cognition 5–6, 8, 11, 22, 36, 216
freediving 14, 93–105, 143, 216
Front-of-House 8, 22–33, 70, 217–8
Fuchs, Thomas 119–20

Galileo 69
games 129, 135–6, 162, 225
See also make-believe

Geertz, Clifford 227–9
genre 39, 160, 225–8
Gibson, James J. 182, 195, 209, 211
glitches 12, 26–7, 29–30, 164–6, 189, 202, 217
See also repair
Globe Theatre 12–13, 24–33, 69–70, 216–18
Goodall, Jane 45
Goodwin, Charles 84
groundlings 24–5, 28–9
Group Theory 156, 160–2, 208

habits, *habitus* 80, 95–7, 143, 179, 194, 215
Hallam, Elizabeth 42
handstands 77–90, 141–4, 216–17, 220
hecklers 29–30
Herndon, Holly 3, 125–38, 145–6, 216, 218
history *See* cultural history, neurohistory, shared history
Hjortborg, Sara Kim 209–11
Hutchins, Edwin 4, 6, 22–3, 38, 102
hypoxia 93–4, 99–100

improvisation 8, 23, 26, 36, 120, 126, 132–3, 148, 159–60, 181, 198–202, 208, 218–19
individuals, individualism 3–6, 9–10, 55, 62–3, 70, 94–5, 102–5, 157–9, 193–5, 217, 226
Ingold, Tim 42, 45
instruction 10–11, 13–14, 83–5, 87, 93–105, 112–18, 142–9
intercorporeality 187–202, 209–10
interdependence 164, 187–9, 209–10
See also collaboration, joint action

interdisciplinarity 5–7, 223–9
  See also disciplines
interventions, artistic 12, 64, 74
intuition 3, 5, 9

jazz 8, 160, 208
joint action, co-action 1, 7–10, 21–3, 144, 157–8, 172–3, 188, 198–202

Kaepernick, Colin 165–6
Kallen, Rachel 207–10, 227
Kimmel, Michael 209–10
kinaesthesia 37, 47, 80, 84–7, 115, 191, 216–17
  See also awareness, bodily
knowledge 11, 22, 80, 99, 102, 111, 115, 121, 142–4, 173–82, 215–17, 224, 228
  See also cognition
Krueger, Joel 3, 145–7

laboratory research *See* experimental psychology
language 65–6, 86–7, 217, 219, 228
Lave, Jean 94, 143
leading and following 70, 117–18, 127, 157–9, 180, 193, 198–9
lighting 24, 161, 217–18
listeners 126–8, 134, 145–6
  See also audience
lockdown *See* COVID–19

make-believe 126–9, 132, 135–6, 145, 218
  See also fiction
Martens, Janno 71, 217, 227
martial arts 2, 109–12, 144–5, 162, 173, 179, 188, 190, 193–4, 196–8, 202, 209–10, 216, 220, 227

materials and craft 56–61, 71
Maxwell, Ian 39, 43–4
McGann, Marek 119–20
memory 115, 182, 218–20
Merleau-Ponty, Maurice 65, 119–20, 144
mind *See* cognition, distributed cognition, 4E cognition
mindful performance 3, 9, 173, 181–3
  See also automaticity, cognition
muay thai 10, 171–83, 209–10
multimodal cognition 38, 81–3, 86–8, 143–4, 219–20
  See also touch
music 2–3, 9, 21, 39, 125–38, 141, 145–8, 158, 160–1, 198, 208, 215–16, 219–20

Nalepka, Patrick 207–8
neuroanthropology 103–4
neurohistory 6
Nguyen, C. Thi 135, 137
non-linear learning 83, 88–9, 116–18, 142
norms and normativity 6–8, 55, 163–4, 183, 210–11
  See also rules
novice learning 79–82, 86–90, 93–105, 114, 142–3, 174–5, 194, 216
nudges 10, 81, 86–8, 97–8, 143, 217
  See also cues, language

Oliveros, Pauline 130, 136
online learning 142–9
oscillation, oscillators 158, 161–2, 207–8
overthinking 3, 71

Packham, Kirsten 40–2, 44
pandemic *See* COVID–19

INDEX

participant observation 11–12, 37, 56, 82, 110–11
  See also apprenticeship learning, ethnography
peer-supported learning 10, 12, 77–9, 85–90, 93–4, 101–4, 142–3, 147–9, 217
performance studies 2, 5–7, 35–6
phenomenology 6, 36–7, 110, 190, 202
physiological breakpoint 97–99
Pini, Sarah 72
Pink, Sarah 45–6
place and 'emplacement' 4, 43–6, 63–5
  See also cognitive ecologies, ecologies of skill, space (theatrical)
plans, planning 9, 57, 115, 129, 172, 177, 180, 182
practical reflexivity 80, 87
probing 11, 172, 177–83
  See also nudges
props *See* equipment
Proto 125, 145–6
push-hands (tai chi) 196–8
push kicks (muay thai) 176–81, 209

Ravn, Susanne 36, 144, 147
Raygun 147–8
rehearsal 21, 127, 215
repair, resilience 12, 183, 200–1
Richardson, Michael 180, 207–8
Rietveld Architecture-Art-Affordances (RAAAF) 12–13, 53–66, 70–2
Rietveld, Erik 71, 217, 227
Rietveld, Ronald 71, 217, 227
Roberts, Tom 3, 145–7, 218
Roepstorff, Andreas 95
rugby union 162, 189–90
Rylance, Mark 26–7, 31
Ryle, Gilbert 211

sandstone 60–1, 71
Sawyer, R. Keith 23, 32
scaffolding 10–12, 22–3, 32, 86–8, 94–103, 113–21, 143
Schneider, Stefan 209–10
Scrabble 216
self-transformation 39, 95, 105, 135, 228
Shakespeare 30–1, 69–70
shared history 2–3, 10, 41, 104, 179
  See also collaboration
sheep herding 207–8
Silva, Pedro 182
Singpayak (Sing) 171–83, 209–11
site-specific artworks 39, 54, 59–60, 63–5
skill 2–5, 8–12, 25–6, 54–8, 79–82, 86–90, 94–5, 103–5, 172–4, 181–3, 217
  See also collaboration
skill acquisition 10–11, 79, 84–9, 93–105, 109–21, 142–9, 215–19
  See also attention, education of
sociology of practice 79–82, 89
sociomaterial systems 3, 10, 54–65, 79–80, 83–8, 102–5
  See also ecologies of skill, equipment
soft assembly *See* assemblages
space, theatrical 24, 28–9
Spawn 3, 125–6, 130–8, 145–6, 218–19
sport 2–3, 8–9, 95, 161–3, 173, 181, 188–9
stacking 191–2
stage manager 22–3, 30
status, status hierarchies 23–7, 30, 70, 96, 112, 115–18, 132, 225
  See also authority
Still Life 58–60
stillness and slow movement 37, 41–3, 72, 117–18

stimulated recall *See* video method
strategies 12, 64, 81–2, 88, 105, 115, 171–83, 195, 197, 200–2, 208–11, 226
symmetry, symmetry breaking 14, 153–66, 180, 196–8, 208
synchrony 120, 158–9, 162–3, 180, 189
synergy 187–202, 209–10, 216, 200, 209–10

tai chi 194, 196–8
Taka-sensei 117–21, 147
Tanaka, Min 36, 39, 45, 48
tango 120, 193–4, 198, 210
teaching *See* instruction
teams, teamwork 9–10, 57, 66, 70, 161–3, 188–90
technology *See* equipment
theatre 8, 13, 21–33
thinking 1–4, 9, 22–3, 32, 71–2, 172–7, 182, 209–11

touch, tactile engagement 81, 87, 143–4
See also multimodal cognition
tourism 31
Tribble, Evelyn 8, 12, 69–70, 101, 216–17
Trollope, Anthony 22
Trusted Strangers 62–4
Turing Test 218–19

urban communities 62–3, 165

video methods 38, 173, 176, 190

Walton, Ashley 160, 208
weather, strange 30, 41, 47
weathering 43–6, 72
Weyl, Hermann 159
Woods, Penelope 25

Yoga 9
See also acroyoga

www.ingramcontent.com/pod-product-compliance
Lightning Source LLC
Chambersburg PA
CBHW062134300426
44115CB00012BA/1912